Color-TV
Training Manual

by

the Howard W. Sams Editorial Staff

Howard W. Sams & Co., Inc.
4300 WEST 62ND ST. INDIANAPOLIS, INDIANA 46268 USA

Preface

When the first edition of this book was introduced more than twenty years ago, its purpose was to provide experienced television technicians with the knowledge necessary to understand and service color-television receivers. The television industry has changed greatly since those early days of color television. Today, color television has become commonplace and the majority of television service requests are for color receivers.

A color television receiver is the most complex piece of electronic equipment usually found in the home. In order to efficiently service a color receiver, the technician must understand the operation and relationships of the circuits that are necessary to reproduce a color picture. In addition, the technician must understand the subject of colorimetry which deals with the art of specifying and identifying colors. This subject is discussed in detail in the first chapter. The next two chapters discuss the requirements and makeup of the color-picture signal.

Section II of this book, which includes Chapters 4 through 9, discusses the fundamentals of each type of circuit used in a color receiver. These circuits include those used in black-and-white receivers and those specifically designed to handle the chrominance portion of the composite color signal. This discussion includes theory of operation for tube-type circuits still in use, as well as the latest solid-state circuits.

The third section of this book covers servicing techniques for color-television receivers. It includes information needed to troubleshoot and adjust color receivers and outlines setup procedures for the various types of color picture tubes now in use.

Color-TV Training Manual has been used by many to gain the knowledge necessary to understand and service color-television receivers. It has been one of the most widely used texts for vocational-training schools and self study. This fourth edition has been updated to include the most-recent developments in color-television circuitry. It should prove invaluable to those just entering the color-television servicing field. The experienced television technician will also find this book a useful reference source.

HOWARD W. SAMS EDITORIAL STAFF

Contents

Section I. Principles of the Color-TV System

CHAPTER 1

CHAPTER 2

CHAPTER 3

Section II. Color Receiver Circuits

CHAPTER 4

CHAPTER 11

Section IV. Appendices

Section I

Principles of the Color-TV System

Chapter 1

Colorimetry, the Science of Determining and Specifying Colors

It is advisable for the service technician to have a basic knowledge of color principles before studying the methods of color reproduction. This is true for the same reason that it is desirable to have an understanding of the nature of radio waves before considering the methods of transmitting and receiving them. For a similar reason, a knowledge of the nature of sound waves and the means by which they are received by the ear is very helpful in the understanding of audio systems. Convinced of the soundness of this reasoning, we are beginning this book with a discussion of color principles before we discuss the technical aspects of the color television system.

The science and practice of determining and specifying colors is referred to as colorimetry. The science of colorimetry has been of great importance to the color television engineers who contributed to the design of the present color television system. Since the system had to be designed to reproduce things as they are seen in nature, the characteristics of light, vision, and color had to be taken into consideration.

Colorimetry is a very complex subject. It is not necessary to be an expert colorimetrist in order to understand the makeup of the color picture signal and the way in which it is utilized in the color receiver; however, a better understanding of the color television system will be attained if some of the most important fundamentals of colorimetry are known. The principles of color as applied to television are slightly different from those that many of us have been taught in connection with other types of color reproduction.

The properties of light and vision should be understood before studying the principles of color. Light is the basis of color, and the eye must be able to convey picture sensation to the brain.

VISION

Human vision is a dual process occurring partly in the eye and partly in the brain. The light output from an object stimulates the eye. This stimulation is transferred to the brain where it is registered as a conscious sensation.

The structure of the eye is similar in many respects to a mechanical instrument. The eye consists essentially of a lens system, a variable diaphragm, and a screen. The variable diaphragm is the iris of the eye, and the screen is the retina. The structure of the eye is shown in Fig. 1-1. Light enters the eye through a transparent layer called the cornea. The amount of light that is allowed to strike the lens is controlled by the contraction and expansion of the iris. During a low light level the iris expands, and during a high light level it contracts. The light passes through the pupil, which is the aperture of the iris, and then through the lens, which is directly behind the iris. There the light is broken up and is focused to form an image on the back wall, or retina, of the eye. The light on the retina stimulates nerve terminals which are called rods and cones. These rods and cones are connected to the brain by a group of nerve fibers called the optic nerve. This nerve furnishes the path by which the light impulses are transferred from the eye to the brain.

The field of vision covers a wide angle of about 200 degrees horizontally and 120 degrees vertically. In the central region of the field of vision, the eye is responsive to color and detail, whereas in the outer region it is chiefly sensitive to motion.

The limits of vision are chiefly determined by four factors: intensity threshold, contrast, visual angle, and time threshold. Intensity threshold is the lowest brightness level that can stimulate the

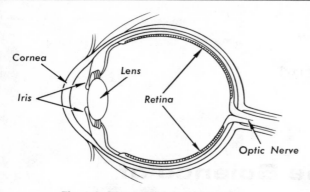

Fig. 1-1. Structure of the human eye.

eye and is very much dependent upon the recent exposure of the eye to light. When a person enters a darkened room, it takes the eye a long time to reach its maximum sensitivity. The required time, which is usually about an hour, differs among individuals. When a person returns to a lighted area, the time it takes for the eye to reach its maximum sensitivity is very short, actually just a matter of minutes.

Contrast represents a difference in the degree of brightness. The limit of vision with respect to contrast is the least brightness difference that can be perceived. The eye is sensitive to percentage changes rather than to absolute changes in intensity.

As an object is made smaller or is placed at a greater distance from the eye, the angle formed by the light rays from the extremities of the object to the eye becomes smaller. This angle is referred to as the visual angle. In order for the eye to respond, the visual angle must be such that the image covers a definite area on the retina. If this area were decreased in size, a point at which the eye could no longer see the object would be reached. This principle is used in eye tests in which the viewer is asked to read the smallest letters possible. Those letters that the viewer cannot read produce on the retina an image that is too small to be useful to the eye. The minimum visual angle is dependent upon the contrast and brightness of the image; for example, an object having sharp contrast could be distinguished at a narrow visual angle while the same size object having a lower contrast might not be visible. The same thing applies to a change in brightness in that a very small object can be more easily seen at a high brightness level than at a low brightness level.

There is a minimum time during which a stimulus must act in order to be effective. This is called the time threshold. If the exposure interval is too short, the rods and cones of the eye do not have time to respond to an image on the retina. The time threshold is also dependent upon the size, brightness, and color of the object.

The following are other factors pertaining to the characteristic of vision:

1. Sensitivity to detail is increased by high contrast, sharp edges, and motion.
2. Straight lines are more readily resolved than curved lines. Horizontal or vertical lines are more easily resolved than diagonal lines.
3. Altering the background of an object changes the appearance of the object. A gray object appears lighter when it is placed on a black background. On the other hand, it appears darker when placed on a white background.

LIGHT SOURCES

The foregoing has shown how the eye is capable of seeing. Let us now discuss the light sources that we must have in order for the eye to be able to use its seeing facilities. In order to see we must have a source of light, just as in the process of hearing we must have a source of sound before we are able to hear. Obviously, if sound waves were not present, nothing would be heard. So, if a source of light were not present, nothing would be seen.

When we speak of light, we usually think of light coming from the sun or light emitted from some electrical lighting. This type of light is referred to as direct light. Another type of light is indirect or reflected light, given off by an object when direct light strikes it. The difference between these two types of light is that indirect light is dependent upon direct light. When light is not shining upon an object, light will not be given off unless the object contains self-luminating properties.

Direct light falling upon an object is either absorbed or reflected. If all of the light is reflected, the object appears white. If the direct light is entirely absorbed, the object appears black. The greater the amount of light reflected by an object, the brighter the object will appear to the eye. In addition, the more intense the direct light source, the brighter the object will become. This can be demonstrated by casting a shadow upon a portion of an object and noting the difference in brightness of the two areas. The portion without a shadow will, of course, appear brighter.

Light is one of the many forms of radiant energy. Any energy that travels by wave motion is considered radiant energy. Classified in this group

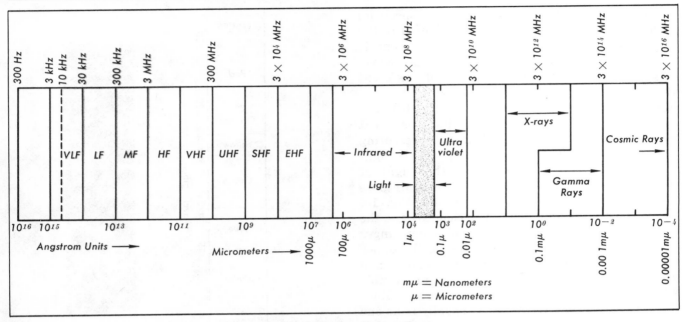

Fig. 1-2. Radiant-energy spectrum.

along with light are sound waves, X-rays, and radio waves.

As shown in Fig. 1-2, light that is useful to the eye occupies only a small portion of the radiant-energy spectrum. Sound is located at the lower end of the spectrum, whereas cosmic rays are at the upper end. Light falls just beyond the middle of the spectrum. Along the top of the spectrum illustrated in Fig. 1-2 is the frequency scale, and along the bottom is the Angstrom-unit scale (10^{-8}

cm). Wavelengths in the region of light may be designated in micrometers (1 micrometer = 10^{-4} cm). These units are also shown along the bottom of Fig. 1-2. Light is made up of that portion of the spectrum between 400 and 700 nanometers.[1]

---

1. Formerly the term *micron* was used for micrometer and *millimicron* for nanometer. The newer terms—micrometer and nanometer—have been adopted by the USA Standards Institute and are synonymous with micron and millimicron, respectively.

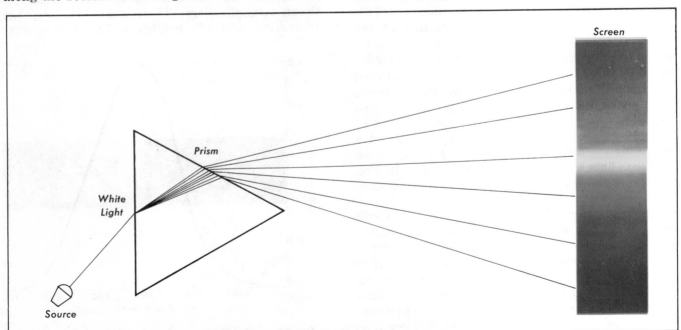

Fig. 1-3. White light dispersed by a prism.

When all wavelengths of the light spectrum from 400 to 700 nanometers are presented to the eye in nearly equal proportions, white light is seen. This white light is made up of various wavelengths which are representative of different colors. This composition can be shown by passing light through a prism, as shown in Fig. 1-3. The light spectrum is broken up into its constituent wavelengths, with each wavelength representing a different color. The ability to disperse the light by a prism stems from the fact that light of shorter wavelengths travels slower through glass than does light of longer wavelengths. Fig. 1-4 shows the relationship of the wavelengths and the colors of the light spectrum. The spectrum ranges from violet on the lower end to red on the upper end. In between fall blue, green, yellow, and orange. A total of six distinct colors are visible when white light is passed through a prism. Since the colors of the spectrum pass gradually from one to the other, the theoretical number of colors becomes infinite. It has been determined that about 125 colors can be identified over the visible gamut. Fig. 1-5 shows the light spectrum in full color.

HOW COLOR APPEARS TO THE EYE

We have shown how light possesses various wavelengths covering the visible spectrum and how the spectrum is divided into various colors. Even though the colors that make up a white light may be of equal intensities, the human eye does not perceive each color with equal efficiency. This fact is due to the physical construction of the eye. It is believed that the cones of the retina respond to color stimuli and that each cone is terminated by three receptors. Each receptor is believed to respond to a different portion of the spectrum, with peaks occurring in the red, green, and blue regions, respectively. An average can be taken of the color response of a number of people, and a standard response for an average person can be derived. This standard response is shown in Fig. 1-5 and is called the luminosity curve for the standard observer.

An inspection of this luminosity curve will show that it is a plot of response versus wavelength. This is the same type of plot that is often used to illustrate the frequency response of a tuned resonant circuit in which maximum response occurs at the resonant frequency and falls off on either side of resonance. With this in mind, we can look at the luminosity curve and see that maximum response occurs at a wavelength of ap-

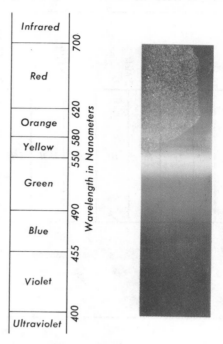

Fig. 1-4. Relationship between colors and wavelengths in the light spectrum.

proximately 555 nanometers and that less response is indicated on either side of that point. From this information, one may see that the average person's eye is most sensitive to light of a yellowish-green color and is less sensitive to blue and red lights.

There are three color attributes that are used to describe any one color or to differentiate between several colors. These are: (1) hue, (2) saturation, and (3) brightnss. Hue is a quality

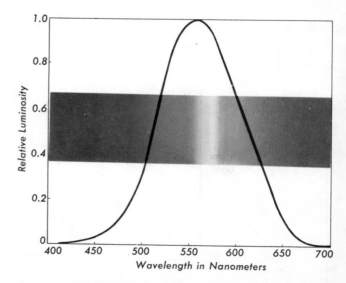

Fig. 1-5. Luminosity response of the eye with respect to the colors of the light spectrum.

used to identify any color under consideration, such as red, blue, or yellow. Saturation is a measure of the absence of dilution by white light and can be expressed with terms such as rich, deep, vivid, or pure. Brightness defines the amount of light energy contained within a given color.

We might consider an analogy between a color and a radiated radio wave. Hue, which defines the wavelength of the color, would be synonymous with frequency, which defines the wavelength of the radio wave. Saturation, which defines the purity of the color, would be synonymous with signal-to-noise ratio, which defines the purity of the radio wave. Brightness, which is governed by the amount of energy in the color, would be synonymous with amplitude, which defines the amount of energy in the radio wave.

Brightness is a characteristic of both white light and color, whereas hue and saturation are characteristics of color only.

Saturation and brightness are often thought of as identical or interrelated qualities of color, whereas they should be considered as separate qualities. It is possible to vary either one of these qualities without changing the other. We might cite an example using an instrument that is familiar to the service technician.

A service oscilloscope presents a green trace on the face of the tube. This trace can be varied in brightness by rotation of the intensity control. An observer can become confused by the fact that the green color appears to be greener at low-intensity levels than at higher intensities. This is often interpreted as an increase in saturation at the low-intensity level. Actually, however, neither the hue nor the saturation of the trace color can change, since these qualities are determined by the chemical properties of the phosphor. The brightness of the trace is the only variable when the intensity control is rotated. The deceiving change in saturation is due to a change in the color response of the eye at low-intensity levels and is often confusing to one who does not know the true reason for this illusion.

In the foregoing example, the brightness level was varied without changing the saturation. Conversely, the saturation of a given color can be changed without varying the brightness, providing certain requirements are met during the process.

In nature, however, a change in saturation is usually accompanied by a change in brightness. This is exemplified by the fact that a pastel color generally appears brighter than a saturated color of the same hue when they are directly lighted by the same source. By changing the lighting on the pastel color (such as by placing it in a shadow), one can decrease the brightness of the pastel color; and it is conceivable that both colors can be made to have the same brightness. Thus, two colors of the same hue but of different saturation can have equal brightness levels.

Any given color within limitations can be reproduced or matched by mixing three primary colors, as will be explained later. This applies to large areas of color only. Color vision for small objects or small areas is much simpler because only two primary colors are needed to produce any hue. This is due to the fact that as the color area is reduced in size it becomes more difficult to differentiate between hues; and for small areas, every hue appears as gray. At this point a change in hue is not apparent; only a change in brightness level can be seen.

As an example, a large area of blue can be readily distinguished from a large area of blue-green; however, when these areas are reduced in size, it becomes more difficult to distinguish between the two colors. Fig. 1-6 illustrates this characteristic when it is viewed at arm's length.

Experiments have been made using sheets of multicolored paper cut in various sizes. A number of things were discovered as these pieces were reduced in size and viewed at a distance.

1. Blues become indistinguishable from grays with equivalent brightness.
2. Yellows also become indistinguishable from grays. In the size range where this happens,

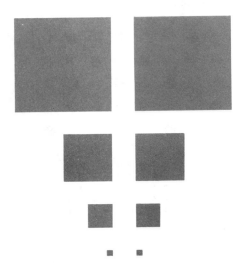

Fig. 1-6. Blocks illustrating good color perception in large areas.

browns are confused with crimson and blues with greens, reds remain clearly distinct from blue-greens, colors with pronounced blue lose blueness, and colors lacking in blue gain blueness.

3. A further decrease in size results in reds merging with grays of equivalent brightness, and blue-greens becoming indistinguishable from grays.

When viewing extremely small objects, the ability of the human eye to identify color is lost and only response to brightness remains. Fig. 1-7 shows clusters of colored dots of three different sizes. Note that a decrease in dot size makes color identification more difficult. (Hold the page at arm's length when viewing this figure.)

Another aspect of color vision is closely related to the persistence of vision exhibited by the human eye. If two objects of different hues are placed side by side and are shuttled back and forth at a sufficiently rapid rate, the two hues will appear as a third hue. This is true for both large and small objects.

COLOR MIXTURE

The production of color may be accomplished by either of two processes. When working with paint pigments, the subtractive process is employed. The other process of mixing colors is called the additive process. This is the process that is employed in color television. These two methods of producing color are rather different. It might be said that the additive process is just the reverse of the subtractive process.

The subtractive process is dependent on incident light. Light falling upon a painted picture is reflected or absorbed. If a certain section of the picture is treated with a red pigment, the light that is reflected is predominantly in the red region of the spectrum and the section will appear red.

The additive process of color mixing used in color television employs colored lights for the production of colors. The colors in the additive process do not depend on an incident light source. Self-luminous properties are characteristic of the additive colors. Phosphorescent signs, which glow in the dark, are good examples of this process. Cathode-ray tubes contain self-luminance properties, so it is only logical that the additive process would be employed in color television.

The three primaries for the additive process of color mixing are red, green, and blue. Two requirements for the primary colors are that each primary must be different and that the combination of any two primaries must not be capable of producing the third. Red, green, and blue were chosen for the additive primaries because they fulfilled these requirements and because it was determined that the greatest number of colors could be matched by the combination of these three colors.

Shown in Fig. 1-8 are the three additive primaries used in color television. Fig. 1-8A shows the primaries as three separate colored lights. Addition of the three colored lights is shown in Fig. 1-8B. When all three primaries are combined in a definite proportion, white is reduced. Red and green combine to make yellow. The combination of red and blue produces magenta (bluish-red), while blue and green combine to make cyan (greenish-blue). Yellow, magenta, and cyan are the secondary colors that are the complements of blue, green, and red, respectively. When a secondary color is combined with its complementary primary, white is produced. For example, combin-

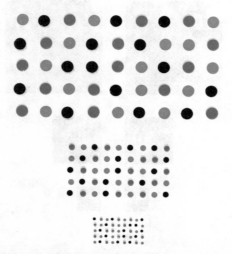

Fig. 1-7. Dot clusters illustrating poor color perception in small areas.

(A) Primaries shown as three separate lights.

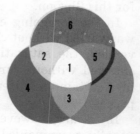

(B) Addition of the three primaries.

Fig. 1-8. Additive primaries.

ing yellow with blue produces white. Cyan added to red results in white, and magenta plus green gives white. Carrying this one step further, the complementary colors when added together produce white. It should be mentioned that specific proportions of these colors must be used in order to produce white.

The foregoing points are shown diagramatically in Fig. 1-9. From this diagram we can see that by mixing colors in certain proportions, we can obtain the following expressions:

$$
\begin{aligned}
\text{Red} &+ \text{Green} = \text{Yellow} \\
\text{Red} &+ \text{Blue} = \text{Magenta} \\
\text{Blue} &+ \text{Green} = \text{Cyan} \\
\text{Yellow} &+ \text{Blue} = \text{White} \\
\text{Cyan} &+ \text{Red} = \text{White} \\
\text{Magenta} &+ \text{Green} = \text{White}
\end{aligned}
$$

Since yellow plus blue equals white, and red plus green equals yellow, then

$$\text{Red} + \text{Green} + \text{Blue} = \text{White}$$

Since the addition of the three primaries can produce white, the addition of the correct proportions of the three complementaries, which are made up of the three primaries, can also produce white. Therefore:

$$\text{Cyan} + \text{Magenta} + \text{Yellow} = \text{White}$$

It is not necessary to overlap the primary colors in the additive process to produce a different color.

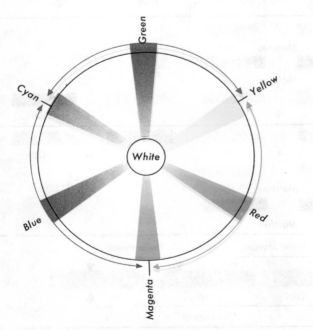

Fig. 1-9. A color circle.

Two sources of colors may be placed in close proximity to each other, and at a certain viewing distance the two colors will blend together and produce the new color. The eye actually performs the additive process. This is referred to as the juxtaposition of color sources. For example, if blue and green are positioned close to each other but not overlapping, the two colors will be blended by the eye and will be seen as cyan when viewed from a distance.

Each additive primary contributes a certain percentage of the brightness in the white which results from mixture. Green is the brightest of the three primaries, red is the second brightest, and blue is the dimmest. This has been determined through experimentation with the response of the eye. The eye responds more to green than to any of the other primary colors. With the total brightness of white considered as unity, green contributes 59 percent of the total, red 30 percent, and blue 11 percent. Therefore, when combining green with red, we have a yellow with a brightness value of 89 percent. Cyan has a brightness of 70 percent. This results from 59 percent of brightness from green and 11 percent of brightness from blue. The third complementary color, magenta, has a brightness of 41 percent. It obtains 30 percent from red and 11 percent from blue. Yellow contains the highest percent of brightness of all of the primary colors and their complements, whereas blue contains the least amount of brightness. The order of brightness for each color is shown in Fig. 1-8B.

There are eight basic rules of colorimetry that apply to the process of color matching. These rules are stated alongside Chart 1-1, which contains an illustration applying to each rule. A knowledge of these rules is very helpful in understanding how matching of colors can be accomplished when using three primary colors.

COLOR SPECIFICATIONS

Standards are necessary in all phases of industry. Imagine the great amount of confusion that would result in the television industry if certain standards of design and construction were not followed. A telecast conforming to one set of transmission specifications could not be received on a receiver designed for reception of a signal having different specifications. Similar difficulties would exist in the specification of colors if standards were not adopted. You can imagine the dismay of someone trying to describe over the telephone the

Chart 1-1.
Color Matching Rules

1.

Any color, with limitations, can be matched by a mixture of three colored lights.

2.

The individual colors which make up a mixture cannot be resolved by the eye.

3.

The total brightness of a mixture is equal to the sum of the individual brightnesses of all colors in the mixture.

4.

If a color match is obtained at one brightness level, the match will be maintained over a wide range of brightness levels. If the brightness of the color to be matched is doubled, a perfect color match will be maintained if the brightness of each color in the matching mixture is doubled.

5.

A color equation can be used to express the formation of a color match. If a color (C) is formed by adding M units of color (M), N units of color (N), and P units of color (P), the resulting mixture can be written:
(C) = M(M) + N(N) + P(P).

6.

Color matches obey the law of addition.
If   (M) = (N)
and   (P) = (Q)
then   (M) + (P) = (N) + (Q).

7.

Color matches obey the law of subtraction.
If   (M) + (P) = (N) + (Q)
and   (P) = (Q)
then   (M) = (N).

8.

Color matches obey the transitive law.
If   (M) = (N)
and   (N) = (P)
then   (M) = (P).

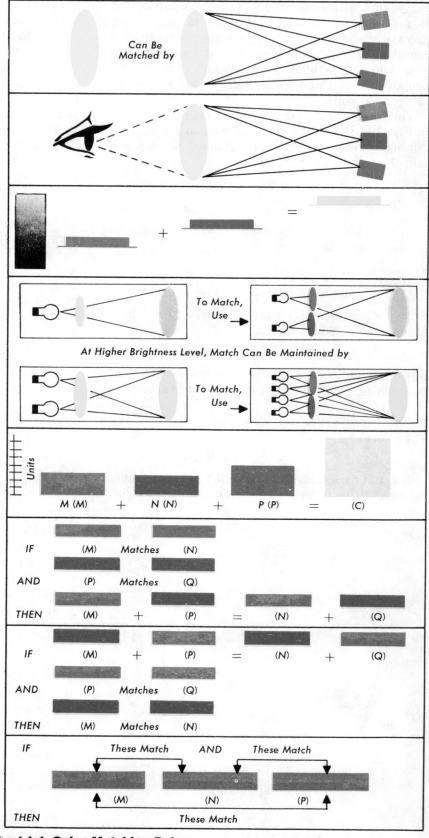

Chart 1-1. Color-Matching Rules

color of a paint that is needed to match a particular color. Such terms as purple, purplish-red, or bluish-purple might be used, but they would certainly not be adequate. The result of such a nonstandardized match would probably be far removed from the original color.

Standards for the specification of color were adopted by the Commission Internationale de l'Eclairage (CIE) at a meeting in 1931. (The English translation of this French name is International Commission on Illumination.) These standards provide that the red primary shall correspond to a light wavelength of 700 nanometers, green to a wavelength of 546.1 nanometers, and blue to a wavelength of 435.8 nanometers.

In the development of a color-matching and specification system, extensive color-matching tests on many observers were conducted using a colorimeter. This is a device that incorporates a photoelectric screen and a series of filters and optical lenses. The method used in these color-matching tests was as follows: one half of the colorimeter screen was illuminated with a spectral hue from a standard source of white light. The hue that was to be matched by the observer was obtained by projecting the light from the standard source through a prism. The hue was selected by moving a plate with a very narrow slit into position so that only the desired hue was allowed to illuminate the colorimeter screen. The other half of the colorimeter screen was then illuminated selectively by the observer with spectral hues of the three additive primaries—red, green, and blue. By the use of independent controls, the energies contributed by each of the primaries were varied by the observer until a color match was thought to be obtained. Each observer was subjected to a series of tests, which constituted attempts to match several selected colors. In order to establish an average response, many persons were used, with each performing similar tests. The results of these tests are known as tristimulus values for color-mixture curves and are illustrated in Fig. 1-10.

Tristimulus values are defined as the amounts of the primaries (red, green, and blue) that must be combined to achieve a color match with all the different colors in the visible spectrum (400 nanometers to 700 nanometers). As an example of how this graph is used, let us select a color having a wavelength of 520 nanometers. The amount of the three spectral primaries needed to match this color can be read from the color-mixture curves. From the graph it can be seen that approximately

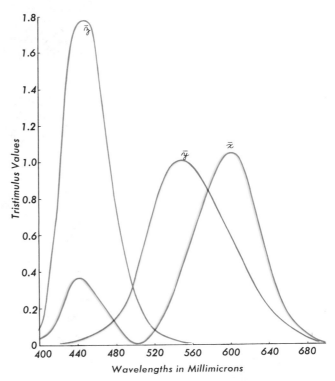

Fig. 1-10. Color-mixture curves.

.09 of red, .09 of blue, and .7 of green are needed to match the color that is specified as having a wavelength of 520 nanometers.

The data contained in the color-mixture curves are not very practical for the specification of all colors. These curves contain information that is required to determine the amounts of the spectral primaries that are needed to match any saturated spectral color. They do not provide information necessary for the matching of desaturated colors; therefore, the need for a more useful means of specifying color is evident.

By the use of mathematical equations, the information contained in the color-mixture curves has been converted to a graphical representation of color on a three-dimensional plane. The conversion equations used to derive the three-dimensional coordinate values x, y, and z are the following:

$$x = \frac{\bar{x}}{\bar{x} + \bar{y} + \bar{z}} \quad (1)$$

$$y = \frac{\bar{y}}{\bar{x} + \bar{y} + \bar{z}} \quad (2)$$

$$z = \frac{\bar{z}}{\bar{x} + \bar{y} + \bar{z}} \quad (3)$$

where,

$\bar{x}$ is a value on the red color-mixture curve,
$\bar{y}$ is a value on the green color-mixture curve,
$\bar{z}$ is a value on the blue color-mixture curve.

By taking values for $\bar{x}$, $\bar{y}$, and $\bar{z}$ from the color-mixture curves at different wavelength intervals and solving for x, y, and z, the results can be used to plot a three-dimensional color diagram. The following are a few examples of these computations:

For a green color of 560 nanometers, the values of $\bar{x}$, $\bar{y}$, and $\bar{z}$ (as taken from the graph shown in Fig. 1-10) are $\bar{x} = .6$, $\bar{y} = 1$, and $\bar{z} = 0$. Substituting these values of x, y, and z into equations 1, 2, and 3, we obtain:

$$x = \frac{.6}{.6 + 1 + 0} = .375$$

$$y = \frac{1}{.6 + 1 + 0} = .625$$

$$z = \frac{0}{.6 + 1 + 0} = 0$$

For a blue color of 480 nanometers, the values of $\bar{x}$, $\bar{y}$, and $\bar{z}$ (as taken from the graph of Fig. 1-10) are $\bar{x} = .1$, $\bar{y} = .15$, and $\bar{z} = .78$. Again substituting these values into equations 1, 2, and 3, we obtain:

$$x = \frac{.1}{.1 + .15 + .78} = .097$$

$$y = \frac{.15}{.1 + .15 + .78} = .146$$

$$z = \frac{.78}{.1 + .15 + .78} = .757$$

For a red light of 600 nanometers, the values of $\bar{x}$, $\bar{y}$, and $\bar{z}$ are $\bar{x} = 1.06$, $\bar{y} = .62$, and $\bar{z} = 0$. Solving, we obtain:

$$x = \frac{1.06}{1.06 + .62 + 0} = .63$$

$$y = \frac{.62}{1.06 + .62 + 0} = .37$$

$$z = \frac{0}{1.06 + .62 + 0} = 0$$

If the procedure is repeated at regular intervals from 400 to 700 nanometers, the results can be used to plot the curve shown in Fig. 1-11. This curve represents color in three dimensions and is referred to as the Maxwell triangle.

The ratios shown in equations 1, 2, and 3 were set up so that the values for x, y, and z at any wavelength would equal unity when added together. This means that at all times x + y + z = 1. This can be checked by adding together the results obtained in each previous example. For instance, in the first example, x = .375, y = .625, and z = 0. It can be seen from this that x + y + z = 1.

Since x + y + z = 1, a similar curve can be shown on a two-dimensional plane, and this curve constitutes a more useful diagram than that shown in Fig. 1-11. Any two of the quantities x, y, and z are sufficient to specify a chromaticity. The third quantity can be found, since x + y + z = 1. By plotting only the values for x and y on the X-Y plane or by projecting the curve of Fig. 1-11 to the X-Y plane, the result is as shown in Fig. 1-12. The curve in this illustration is called the *CIE chromaticity diagram*. The projection shown in Fig. 1-12 was accomplished by photographing the structure shown in Fig. 1-11 from above and along the Z axis. This eliminated the dimension on the Z axis and provided a proportionate diagram having only X and Y axes. The two-dimension CIE chromaticity diagram shown in Fig. 1-12 is used extensively in the specification of color, since all colors contained within the locus of the CIA diagram can be specified in terms of X and Y.

If we examine the diagram in Fig. 1-12, we see that the horseshoe curve, which is known as the spectrum locus, is graduated into numerals ranging from 400 in the left-hand corner to 700 at the extreme right. These figures represent the wavelengths of the various colors in nanometers: the blues (including violet) extend from approximately 400 to 490 nanometers, the greens extend from approximately 490 to 550 nanometers, the yellows extend from approximately 550 to 580 nanometers, and the reds (including orange) extend from approximately 580 to 700 nanometers.

Any point that is not actually on the spectrum locus but that lies within this diagram can be defined as some mixture of spectrum colors. Since white is such a mixture, it naturally falls within this area.

It might be well to point out that there is another method of specifying colors near the white region. This method uses the temperature in kelvin to designate a particular color. (Kelvin temperature equals degrees in Celsius plus 273.) All of us have seen metal heated to various temperatures, and we have seen how it will change in color from dull cherry red to white as the temperature is increased. If we increase the temperature still further, the metal will take on a decidedly bluish cast. These colors can be shown on the chromaticity diagram if we draw an imaginary arc starting in the reddish region on the right side of the diagram and extending up through the orange into the white area and down toward the bluish region to the left. Such facts are of importance when we consider the three standard sources of

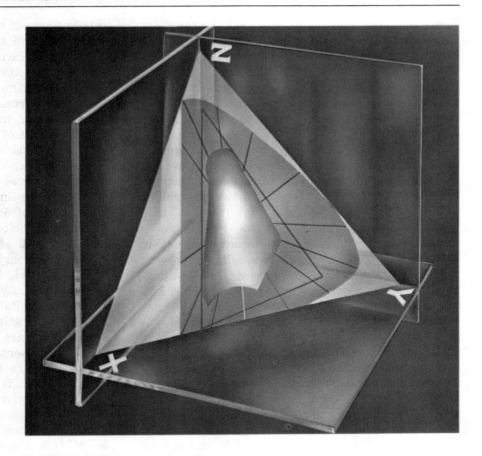

Fig. 1-11. The Maxwell triangle. By using the proper equations, the values obtained from the color-mixture curves can be converted to the values of x, y, and z. These values can then be plotted in terms of X, Y, and Z. The three-dimensional curve shown here will result.

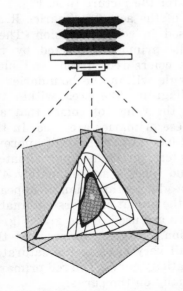

Fig. 1-12. The CIE chromaticity diagram and NTSC triangle. This diagram is a result of the projection on the X-Y plane of the three-dimensional curve shown in Fig. 1-11. This projection can be accomplished by photographing the structure with the camera in line with the Z axis. To visualize how this is done, assume the camera position to be directly above the Z axis.

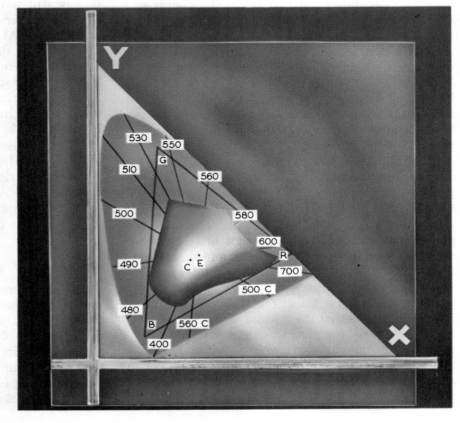

light and how they were specified by the CIE. These three standard sources of light were chosen for measurement purposes, and they are known as illuminant A, illuminant B, and illuminant C.

Illuminant A was selected as a match in color to a conventional tungsten lamp. To achieve this, a tungesten filament was heated to approximately 2500 kelvin.

Illuminant B was selected to give an approximate match to direct sunlight and was achieved in the same manner as illuminant A, except that the tungsten filament was used with two prescribed liquid filters to give a color that matches 4800 kelvin.

Illuminant C gives a radiation that most nearly matches daylight. Again, tungsten filaments were used with the filter solutions arranged so that they would produce a color of approximately 6500 kelvin. Illuminant C is the only one we need be concerned with. Because it is considered the most satisfactory from a viewing standpoint, it is the one that has been selected by the NTSC as the reference white for color television work. It is shown in the chromaticity diagram as point C.

With reference to white, one other term which we may encounter and with which we should be familiar is *equal energy white*. This is shown on the chromaticity diagram as point E and can be described as white composed of equal amounts of energy from the three primary colors of red, green, and blue.

Since saturation of color is defined as the degree of freedom from white, the spectrum colors that lie directly on the horseshoe curve can be said to be 100 percent saturated because they contain zero amount of white. At point E, only white light is present; so it can be said to be zero percent saturated. Various percentages of saturation fall along a straight line drawn between any point on the spectrum locus and point E. As we move toward point E, the saturation will be decreased. Conversely, as we move toward the curve, the saturation will be increased. Thus, we can see that a 100 percent saturated color is one that has 100 percent purity or freedom from white, and that a desaturated color is a color that contains some amount of white light.

Referring again to the chromaticity diagram in Fig. 1-12, we see that the bottom of the horseshoe curve has been completed with a straight line drawn from purplish-blue to red. Although this line completes the curve, it should not be considered in the same sense as the rest of the horseshoe. The reason for this is that the colors along

this line cannot be assigned dominant wavelengths within the limits of the spectrum; therefore, these colors are known as nonspectral colors. They can, however, be expressed as the complements of some of the spectrum colors that fall directly on the horseshoe curve. A line from 500 nanometers has been extended through white to the straight line, and the point of intersection is labeled 500C; or, in other words, the nonspectral color at the point is the complementary color of the bluish-green of a wavelength of 500 nanometers. The same thing is true of the line that has been extended from a wavelength of 560 nanometers; 560C is the complementary color of a yellowish-green with a wavelength of 560 nanometers.

When primary colors were selected for color television work, it was found that those primaries must of necessity be limited by the color phosphors that were available for the picture tube. Fig. 1-12 shows the location of the actual primaries R, G, and B that are used in color television. These points represent the primaries selected by the NTSC and are the colors red, green, and blue. They define a triangle within the boundaries of the chromaticity diagram; the area within the triangle represents the range of colors that are obtainable when these primaries are used. In the NTSC triangle, red has a wavelength of approximately 610 nanometers, green is approximately 540 nanometers, and blue is approximately 470 nanometers. At first glance, this triangle appears much smaller than the gamut of colors obtainable when ideal primaries are used. If we give Fig. 1-12 a closer inspection, however, we see that the NTSC primaries fall very close to the saturated colors on the chromaticity curve. The red primary, for example, is actually on the curve.

Fig. 1-12 also shows the colors obtainable from modern printing inks. It is apparent that the NTSC color triangle covers a considerably larger area than the area of the printing inks. It would seem from this that the colors that can be displayed in color television are entirely adequate.

SUMMARY

The characteristics of human vision are important in that the eye is the instrument that judges the quality of color reception. Human vision is a remarkable function, but it has certain deficiencies and irregularities which limit its effectiveness. An understanding of these limitations is extremely helpful to anyone engaged in color television work. Examples of these limitations are:

the luminosity response of the eye to various colors, intensity and time thresholds, contrast limitations, and the visual-angle requirements. It was also pointed out that certain illusions occur particularly with respect to brightness and saturation changes.

The two types of light sources, direct and indirect, have been explained. In addition, the constituent colors and wavelengths of the light spectrum have been investigated. The attributes of hue, saturation, and brightness were described; and their importance will become increasingly evident as we proceed into the study of the color television system.

The development of the chromaticity diagram was covered in some detail. There were two major reasons for this: (1) the diagram is based on actual tests of human vision, and this fact lends authority to the data presented, and (2) the NTSC triangle, which is used in servicing work, is based to a great extent on the chromaticity diagram.

Since in the color television system the desired colors are produced by a mixing action, a thorough coverage of color-mixture rules has been presented. These rules are basic, and a knowledge of them will prove helpful in analyzing and adjusting color receivers.

QUESTIONS

1. If a small area of red direct light is placed next to a small area of green direct light, what color is seen by the eye?
2. What region of the visible spectrum appears brightest to the eye? Why?
3. What is meant by desaturation?
4. What are the relationships of all colors that lie on a line drawn on the chromaticity diagram from point E to any point on the spectrum locus?
5. What are the three NTSC primaries? What are the three secondary colors, and how are they produced?

Chapter 2

Requirements of the
Composite Color Signal

The composite color signal has a number of requirements; not only must it carry color information, but it must also be compatible with the long-established system of monochrome (black-and-white) television. In other words, the signal must be such that it can be received on a monochrome receiver in black and white without any modification of the receiver. In addition, the part of the signal that conveys color must be transmitted in such a manner that it does not appreciably affect the quality or type of picture reproduced by the monochrome receiver, which is tuned to the color signal.

The signal must represent the scene according to its color, and the colors must be transmitted in terms of the three chosen primaries—red, green, and blue. By some means, the three physical aspects of brightness, hue, and saturation must be conveyed by the signal for each color in the scene, because the eye sees color in terms of these aspects.

In order to make the color system compatible, the specifications of the standard monochrome signal had to be retained. This meant that such things as the channel width of 6 megahertz (MHz), the aspect ratio of 4 to 3, the number of scanning lines per frame at 525, the horizontal-scanning and the vertical-scanning rates at 15,750 Hz and 60 Hz, respectively, and the video bandwidth of 4.25 MHz had to remain the same within narrow tolerances. To these basic specifications, provisions had to be added to convey the color elements by means of a signal that will hereafter be known as the chrominance signal.

Even if the same specifications were retained, the color system would not be compatible if the composite color signal did not contain a signal that would convey brightness. To satisfy this requirement, a signal that is representative of the brightness of the colors in the scene must be transmitted together with the chrominance signal. This brightness signal is very much the same as the video signal used in standard monochrome transmission, and it will be referred to hereafter as the luminance signal. It is transmitted by amplitude modulation of the picture carrier in such a manner that an increase in brightness corresponds to a decrease in the amplitude of the carrier envelope.

Putting a chrominance signal in the allotted channel of 6 MHz created a difficult problem, since this chrominance signal had to be transmitted along with the luminance signal and had to be included without objectionable interference to the luminance signal. This was accomplished by proper placement of the chrominance signal within the band of video frequencies and by limitation of the bandwidth of this signal.

The chrominance and luminance signals are included within the 4.25-MHz video band by an interleaving process. This process is possible because the energy of the luminance signal concentrates at specific intervals in the frequency spectrum. The spaces between these intervals are relatively void of energy, and the energy of the chrominance signal can be caused to concentrate in these spaces. A detailed discussion of the interleaving process is presented later in this chapter.

The chrominance signal is conveyed by means of a subcarrier. The frequency of this subcarrier was chosen so that its energy would interleave with the energy of the luminance signal. The energy of each of these signals is conveyed by the video carrier. The subcarrier frequency is high enough in the video band that the subcarrier sidebands, when they are limited to a certain bandwidth, do not interfere with the reproduction of the luminance signal by a monochrome receiver. The choice of frequency for the chrominance subcarrier and the development of the chrominance subcarrier will be covered in detail.

The chrominance signal must convey energy that represents the primary colors. Color could be transmitted by three separate signals, each representing a primary color, but a channel width of at least 12.75 MHz would then be required. Since this would take away the idea of compatibility, color had to be represented in some other manner in order to utilize the standard 6-MHz channel width.

Three signals—the red, the green, and the blue—are obtained from the camera. Portions of these three signals are used to form the luminance signal. This leaves three signals, which are referred to as color-difference signals. Two of these are proportionately mixed together to form two other signals that are used to modulate the chrominance subcarrier.

A method of modulation known as divided-carrier modulation may be employed in order to place two different signals upon the same carrier. The subcarrier is effectively split into two parts, and each portion of the subcarrier is modulated separately. Then the two portions are combined to form the resultant chrominance signal. The amplitude and phase of this signal vary in accordance with variations in the modulating signals. A change in amplitude of the chrominance signal represents a change in color saturation, and a change in phase represents a change in hue.

A reference signal of the same frequency as the subcarrier frequency is transmitted in the composite color signal. This reference signal, called the color burst, has a fixed phase angle and is employed by the color receiver in order to detect properly the colors represented by the chrominance signal.

The foregoing discussion stated that the composite color signal contains a luminance signal and a chrominance signal. A color-burst signal is also transmitted, along with the conventional blanking and sweep-synchronizing signals. Let us now examine in greater detail the methods employed in making up the composite color signal.

THE INTERLEAVING PROCESS

The interleaving process, as mentioned previously, makes it possible to transmit the composite color signal within a channel no wider than that used for monochrome transmission. The use of this process was a great step toward formulating the compatible color system that was adopted by the Federal Communications Commission (FCC) for use in the United States.

It was not known how to make the color system compatible until a discovery by Pierre Mertz and Frank Gray was taken into consideration.[1] Their studies concerning the scanning process used in telephotography and television proved that the energy produced by scanning an image concentrates at specific intervals in the frequency spectrum. It was further shown that these points of energy concentration occur at frequencies computed as whole multiples of the scanning rate. The actual proof of this phenomenon involves a series of complicated mathematical formulas not practical to reproduce here, but a more or less general idea of the reasons for such energy concentration is presented in order to help the reader to understand the interleaving process.

A mathematical solution would show that any video signal produced by scanning an image contains an infinite number of pure sine waves. As an example, waveform E in Fig. 2-1 can be matched by combining waveforms A, B, and C. If the eighth harmonic of waveform A is added to the frequency of waveform E, waveform F will result. Note that the frequency of each of the pure sine waves is either the fundamental or a harmonic of the recurring rate of waveforms E and F. Although more complex than waveform E or F, every video waveform contains an infinite number of pure sine waves, each of which has a fundamental or a harmonic frequency of the recurring rate of the waveform.

The monochrome video waveform shown in Fig. 2-2 contains an infinite number of sine waves which have frequencies that are multiples of the fundamental frequency of the entire waveform. Because this fundamental frequency is determined by the rate of scanning, a video waveform produced by scanning an image at a constant rate will contain an infinite number of sine waves which have frequencies that are harmonics of the line-scanning frequency. If the waveform shown in this figure were followed through several successive scanning lines, there would be very little change in its composition. Thus, the waveform recurs at the line-scanning frequency.

1. Pierre Mertz and Frank Gray, "A Theory of Scanning and Its Relation to the Characteristics of the Transmitted Signal in Telephotography and Television," The Bell System Technical Journal, Vol. XIII, No. 3, July 1934.

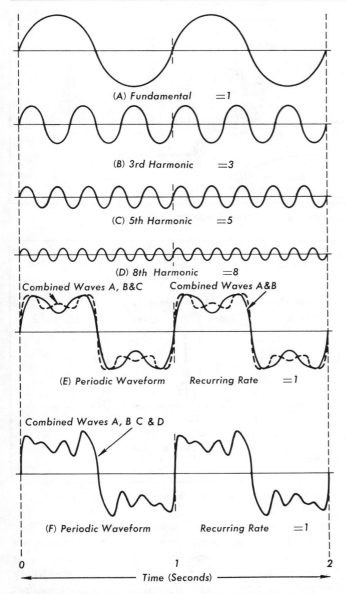

(A) Fundamental   =1

(B) 3rd Harmonic   =3

(C) 5th Harmonic   =5

(D) 8th Harmonic   =8

Combined Waves A, B&C   Combined Waves A&B

(E) Periodic Waveform   Recurring Rate   =1

Combined Waves A, B C & D

(F) Periodic Waveform   Recurring Rate   =1

0   1   2

Time (Seconds)

Fig. 2-1. Sine-wave components of a periodic waveform.

trations of energy occur in the video spectrum at whole multiples of the frame and line frequencies. More energy concentrates at multiples of the line-scanning rate than at multiples of the frame rate because of the greater number of successive waveforms at the line frequency.

Present-day monochrome transmission follows this principle of energy distribution. The drawing presented in Fig. 2-3 shows the frequencies at which the concentrations of energy occur. It may be noted that nearly half the video spectrum is unused.

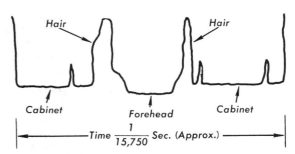

Hair   Hair

Cabinet   Forehead   Cabinet

Time $\frac{1}{15,750}$ Sec. (Approx.)

Fig. 2-2. Complex waveform of video voltage produced by scanning one line in the televised scene.

It follows that for an image which is not moving rapidly, the waveform of a particular line will be repeated during the scanning of each successive frame; therefore, the waveform also recurs at the frame rate. As a result of repeated scanning, a number of recurring waveforms at the frame frequency and at the line-scanning frequency are produced.

Since the transmitted signal comprises recurring waveforms at the frame and line frequencies, and since these waveforms contain an infinite number of waves having frequencies that are harmonics of the frame and line frequencies, the frequency spectrum contains a concentration of energy at each harmonic. This means that concen-

Since the scanning rates for the chrominance signal and the luminance signal are the same, the concentrations of energy produced by both are spaced at the same intervals. It is feasible, therefore, that the bands of concentrated energy of the chrominance signal could be spaced between the bands of the luminance signal. As seen in Fig. 2-4, the spaces in the frequency spectrum occur at odd multiples of one-half the line frequency. If a subcarrier frequency equal to an odd multiple of one-half the line-scanning frequency is chosen, the chrominance and luminance signals will be interleaved.

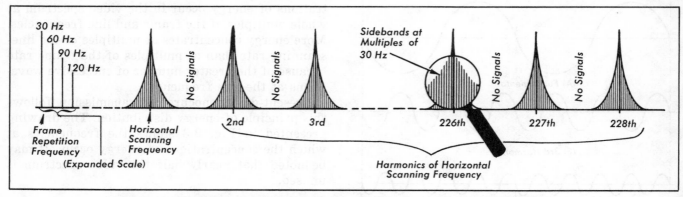

Fig. 2-3. Distribution of energy in the frequency spectrum of a standard monochrome signal.

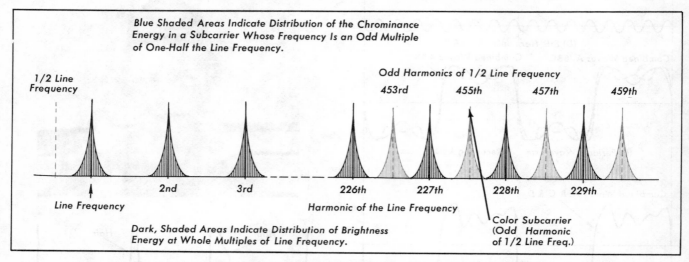

Fig. 2-4. The interleaving of brightness and color signals in the frequency spectrum.

DEVELOPMENT OF THE SUBCARRIER FREQUENCY

Most of us are familiar with the nature of a carrier signal. It is a signal in which some feature—such as amplitude, frequency, or phase—may be made to vary in accordance with the characteristics of a modulating signal.

In radio transmission and in monochrome television, the modulation can be recovered by a detection process. In color television, however, a portion of the modulation on the video carrier performs by itself the functions of a carrier. This modulating frequency is called a subcarrier. After being detected and separated from the video carrier, the subcarrier must undergo further demodulation before the characteristics of the modulating signal, which is conveyed by the subcarrier can be obtained.

The frequency of the chrominance subcarrier is governed by several factors. The subcarrier frequency must be high enough above the video carrier to keep interference at a minimum in either monochrome or color receivers. If a frequency near those in the upper limit of the video range is used, it will undergo attenuation in the relatively narrow bandpass circuits of a monochrome receiver. Moreover, if this frequency does reach the picture tube, the dot structure it produces will be very fine and not too objectionable.

On the other hand, the subcarrier frequency must be low enough that its upper sidebands will fall within the established range of the video frequencies. Specifications for the color signal indicate that the upper limit of the subcarrier sidebands should be 0.6 MHz above the frequency of the subcarrier. It follows that a frequency as high as 3.6 MHz can be used as the subcarrier frequency if we consider that the practical video bandwidth for color transmitters and receivers has been established as approximately 4.2 MHz.

It has been shown that the signal energy produced by scanning an image will concentrate around the harmonics of the scanning frequencies.

Nearly half the space in the video spectrum is therefore left unused. The location of the unused space is another determining factor for the subcarrier frequency because the harmonics of the subcarrier must concentrate in this space. A frequency that is an odd multiple of one-half the line frequency must be used for the chrominance signal so that the interleaving process will take place. A tentative subcarrier frequency f_s can be computed as:

$$f_s = \frac{15,750 \times 10^{-6} \times 455}{2}$$
$$= 3.583125 \text{ MHz}$$

The multiple 455 is used to keep the frequency close to 3.6 MHz above the video carrier.

This 3.583125-MHz frequency was not adopted for the color-transmission standards because of an objectionable feature which may be described as follows. Monochrome receivers that employ an intercarrier type of sound system develop a 4.5-MHz signal at the output of the video detector. When this 4.5-MHz signal beats with the color subcarrier, a difference of approximately 900 kHz is produced. Experiments were conducted, and it was found that when the 3.583125-MHz frequency was used for the color subcarrier, the beat frequency created a distracting pattern on the screen of a monochrome picture tube. Further experimentation proved that the beat frequency would be less objectionable if it were an odd multiple of one-half the line frequency.

Naturally, it would have been impractical to change the 4.5-MHz intercarrier frequency which is accepted as standard in so many existing receivers. This made it necessary to select a slightly lower frequency for the color subcarrier than the tentative value given; consequently, slightly lower line and field frequencies had to be used in order to retain their frequency relationship to the subcarrier frequency.

The difference frequency between the video carrier and the subcarrier and the difference frequency between the subcarrier and the sound carrier must both be taken into consideration when the new line and field frequencies are being determined. For the best results, it is desirable that each of these two difference frequencies should be some odd multiple of one-half the line rate. Since the addition of two odd harmonics of one-half the line rate will yield an even multiple of the line rate, the separation between the video carrier and the sound carrier must be defined as an even multiple of the new line frequency. Using a 15,750-Hz

line rate as the basis for determining this multiple, it was found that the 286th harmonic would be at a frequency of 4.504500 MHz.

This new line frequency, f_L, can be computed as follows:

$$f_L = \frac{4.5 \times 10^6}{286} = 15,734.264 \text{ Hz}$$

Thus, the 286th harmonic of the new line frequency is equal to the 4.5-MHz picture-to-sound separation.

Since each frame must consist of 525 lines, it follows that the new field frequency, f_F, can be computed as follows:

$$f_F = \frac{f_L}{525} \times 2 = 59.94 \text{ Hz}$$

Now the subcarrier frequency, f_S, becomes:

$$f_S = \frac{455}{2} \times f_L = 3.579545 \text{ MHz}$$

Note that the new scanning frequencies used for color transmission are slightly below the nominal values used in monochrome receivers; however, the changes amount to less than the one-percent tolerance allowed, and the new frequencies will fulfill the requirements for compatibility in monochrome reception.

For purposes of maintaining close synchronization of color receivers, the tolerance for the subcarrier frequency is set at ±.0003 percent, or around ±10 Hz; and the rate of change cannot be more than 1/10 Hz per second. The same percentage of tolerance applies to the field and line frequencies. In actual practice, all of these frequencies are developed from the same source in such a manner as to minimize variations from the established figures.

DIVIDED-CARRIER MODULATION

It has been pointed out that for purposes of color transmission, a chrominance signal is required; moreover, the chrominance signal must represent two color signals that are separable from each other. One subcarrier at a frequency of 3.579545 MHz above the picture carrier is available to convey both color signals. Consequently, some method of modulating one carrier with two signals must be utilized at the transmitter.

As can be seen in Fig. 2-5, a fundamental block diagram illustrates the manner in which one carrier may be modulated by two signals. A subcarrier generator produces a sine wave of constant

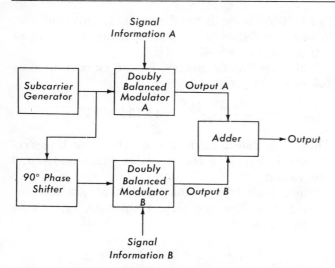

Fig. 2-5. Block diagram of divided-carrier modulation system.

frequency and amplitude. This subcarrier is then applied to two doubly balanced modulator circuits represented by blocks A and B. The subcarrier coupled to the modulator in block B has been subjected to a 90-degree phase shift. One of the two modulating signals is applied to modulator A, and the other is applied to modulator B.

In order to show what occurs in each of the balanced modulator circuits, a simple schematic diagram representing the modulator in block A is presented in Fig. 2-6. The modulating signal is passed through a phase-splitter circuit to produce two signals of opposite polarity. This causes the signal at the grid of tube No. 1 to be 180 degrees out of phase with the signal at the grid of tube No. 2. The subcarrier also goes through a phase-splitting process as a result of the transformer

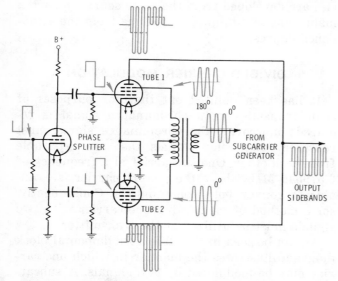

Fig. 2-6. Schematic diagram of doubly balanced modulator.

action, and the suppressor grids are fed with signals of equal amplitude but of opposite polarity.

When the modulating signal at the grid of tube No. 1 is in the positive portion of its cycle, a subcarrier signal of increased amplitude is produced. As the signal at the grid goes through the negative portion of its cycle, a subcarrier signal of lower amplitude appears at the output. The waveform of the subcarrier signal is shown in the figure at the plate of tube No. 1. A similar operation takes place in tube No. 2, except that the modulating signal goes through its negative cycle first. As a result, the output signal having the low amplitude is produced before that having the high amplitude. The resultant waveform is shown in the figure at the plate of tube No. 2.

The operation of the entire circuit is dependent upon the fact that the plate circuits of tubes Nos. 1 and 2 are tied together. The waveforms shown at the plates do not exist separately but are actually combined because of the common plate circuit. By noting in the illustration of the two waveforms shown, one at the plate of each tube, it can be seen that the signals are 180 degrees out of phase. Because the plates are tied together, the output sidebands will consist of a combination of the output waveforms produced by the two tubes. Since these output waveforms are always 180 degrees out of phase, a cancellation effect occurs. The amplitude of the total output becomes the difference in the amplitudes of the two signals. In addition, the phase of the output sidebands agrees with the phase of the signal having the greatest amplitude.

It can be seen that if the signal applied to the grid of the phase splitter is increased in amplitude, then the amplitude of the output signal will also increase. The phase of the output sidebands will change 180 degrees when the polarity of the modulating signal is reversed; consequently, the output signal represents the modulating signal in terms of the phase and amplitude of the subcarrier frequency. If no signal is applied to the grids of tubes Nos. 1 and 2, no signal will appear in the output because both tubes will conduct equally and complete cancellation will result.

In Fig. 2-5, there are two blocks representing doubly balanced modulators. The modulator in block B operates in the same manner as that described for the one in block A, with the exception that the subcarrier input is delayed 90 degrees. This delay causes the output of modulator B to be displaced 90 degrees in phase with reference to the output of modulator A. In this respect, it should be remembered that the output of modulator B

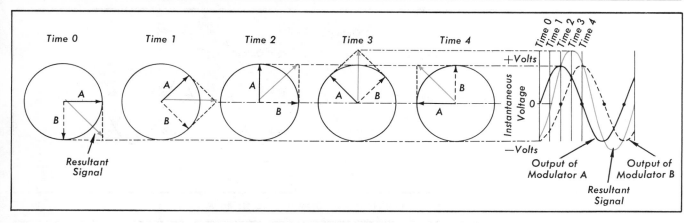

Fig. 2-7. Vector representations of instantaneous voltages at various times.

can either lead or lag that of modulator A by 90 degrees, depending on the polarity of the modulating signal introduced into each balanced modulator circuit.

Consider a particular case in which the output of modulator A is equal in amplitude to the output of modulator B but leads the latter output by 90 degrees. See Fig. 2-7. Since the values of both of these signals are continuously changing, and since the signals are displaced in phase, it is easier to analyze their relationship through the use of vectors. Thus, we can consider both signals as vectors rotating at the same speed but with one leading the other by 90 degrees.

The first pair of vectors in Fig. 2-7 is shown at zero time. The vector representing the instantaneous value of A leads that representing the instantaneous value of B of 90 degrees because the motion of vectors is considered to be counterclockwise. At zero time, the voltage of A is zero and the voltage of B is at a maximum negative value. The resultant signal at zero time is equal to the value of B. This can be seen in the vector diagram as well as in the sine-wave representation of the signal at time zero.

If we allow the vectors to rotate another 45 degrees, they will appear as shown in the vector diagram for time No. 1 in Fig. 2-7. Note that at time No. 1, the vector of A has reached an amplitude of 70.7 percent of the positive peak and the vector of B is 70.7 percent of the negative peak. The resultant signal has an amplitude of zero. In the sine-wave diagram, these values are shown at time No. 1. The positions of the vectors at times 2, 3, and 4 indicate that the instantaneous voltages in the vector diagrams are equal to those on the sine-wave diagram at their respective times.

If the rotating vectors in Fig. 2-7 were analyzed at every degree throughout their complete cycle,

the waveforms that are represented by these vectors could be traced as they have been in the sine-wave diagram. The length of each vector equals the peak amplitude of the signal it represents.

In the foregoing discussion, the outputs of modulators A and B were considered to be of equal amplitude, and they produced a resultant signal shown by the blue line in Fig. 2-8A. The associated vector diagram is drawn for the vector positions at zero time. In actual practice, the output from each of the doubly balanced modulators will vary in amplitude and undergo a 180-degree phase shift from time to time. As an illustration of this condition, Fig. 2-8B shows the resultant signal produced when the output of B is one-fourth the output of A. Notice that the amplitude of the resultant signal is less than its amplitude in Fig. 2-8A. In addition, a phase shift in the resultant signal is indicated by the fact that in Fig. 2-8A the resultant signal lags the output of A by 45 degrees, whereas in Fig. 2-8B it lags by only 14 degrees.

The voltages from the outputs of modulators A and B are combined in the adder stage (see Fig. 2-5). The output of this stage is a single waveform which varies in amplitude and phase in accordance with the amplitude and phase of each of the two signals introduced to modulators A and B. Thus, two modulating signals are impressed upon a single subcarrier, and these two signals can be recovered by reversing the modulation process at the receiving end.

COLOR SYNCHRONIZATION

The chrominance signal changes phase with every change in the hue of the color it represents, and the phase difference between the chrominance signal and the output of the subcarrier generator identifies the particular hue at that instant. When

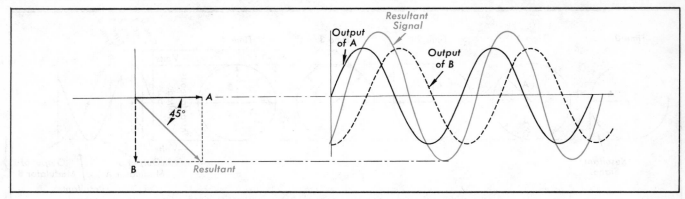

(A) Amplitudes of outputs of modulators A and B are equal.

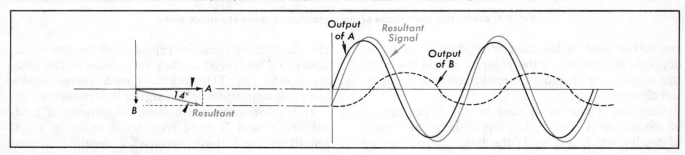

(B) The output of modulator B has one-fourth the amplitude of the output of modulator A.

Fig. 2-8. Voltages resulting from the addition of two output sidebands separated in phase by 90 degrees.

the chrominance signal reaches the receiver, the receiver must have some means of comparing the phase of the signal with a fixed reference phase which is identical to that of the subcarrier generator at the transmitter. This reference phase is provided in the receiver by a local oscillator that is synchronized with the subcarrier generator by means of a color-burst signal transmitted during the horizontal-blanking period. The color burst consists of a minimum of eight cycles at 3.579545 MHz.

As shown in Fig. 2-9, the color burst is placed on the back porch of the horizontal-blanking pedestal. When located at this point, the burst will not affect the operation of the horizontal-oscillator circuits because the horizontal systems used in existing receivers are designed to be immune to any noise or pulse for a short time after they have been triggered. Since the average voltage of the color burst is the same as the voltage of the blanking level, the burst signal will not produce spurious light on the picture tube during the retrace period. Fig. 2-10 is a waveform of the burst signal applied to a wideband oscilloscope which was synchronized at the horizontal-scanning frequency.

A color receiver is designed to extract the color burst from the transmitted signal. This reference signal is used to synchronize the color section of

the receiver in much the same manner as the horizontal and vertical pulses are used to synchronize the horizontal- and vertical-sweep sections.

CHROMINANCE-CANCELLATION EFFECT

From the standpoint of compatibility, how does color transmission affect monochrome reception? The frequency of the color subcarrier was established at 3.579545 MHz. The sidebands extend 0.5 MHz above and 1.5 MHz below this frequency. Fig. 2-11 shows the positions that the luminance and chrominance signals occupy in the video spectrum. Note that the chrominance signal falls within the bandpass limits of existing monochrome receivers, which ordinarily have good frequency response to approximately 3.5 MHz.

Because the chrominance signal falls within these limits, a dot pattern will be reproduced on the screen of a monochrome receiver which is tuned to a composite color signal. The dot structure of this pattern is very fine because the frequency of the chrominance signal is relatively high in the video spectrum. In addition, the color subcarrier was set at a frequency that is an odd harmonic of one-half the line-scanning rate, and the brightness variations produced on the screen by such a frequency go through a cancellation effect.

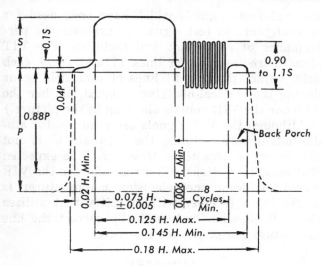

NOTES

1. The burst frequency shall be the frequency specified for the chrominace subcarrier. The tolerance on the frequency shall be ±0.0003% with a maximum rate of change of frequency not to exceed 1/10 hertz per second.

2. The horizontal scanning frequency shall be $\frac{2}{455}$ times the burst frequency.

3. Burst follows each horizontal pulse, but is omitted following the equalizing pulses and during the broad vertical pulses.

4. The dimensions specified for the burst determine the times of starting and stopping the burst, but not its phase.

5. Dimension "P" represents the peak-to-peak excursion of the luminance signal, but does not include the chrominance signal.

Fig. 2-9. Specifications for the color burst.

The explanation of this phenomenon is given in the following discussion.

It has been shown that video signals can be broken down by analysis into many sine waves; thus, the brightness variations produced along any given scanning line by each of these waves is of a sinusoidal nature. Since it has been established that all of the sine-wave components of the luminance signal are harmonics of the line and frame frequencies, the brightness variations produced by these components go through a whole number of cycles during the scanning of any given line or frame. This means that either in the next line or in the next frame, the variations recur in phase.

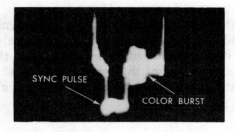

Fig. 2-10. Waveform showing the color burst and horizontal sync pulse in the composite color signal.

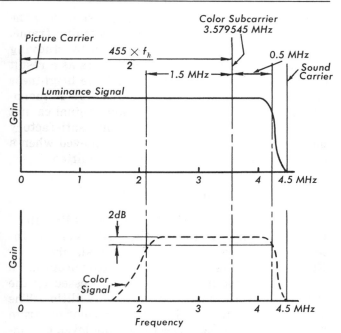

Fig. 2-11. Complete video spectrum of a standard color transmission.

A reinforcing effect is produced and is like that illustrated in Fig. 2-12A.

In the case of the chrominance signal, however, the opposite condition exists. The frequency of the chrominance signal is an odd harmonic of one-half the line frequency. During the scanning time of any one line on the picture tube of a monochrome receiver, the chrominance signal goes through a certain number of cycles plus a half cycle. During the scanning of the succeeding line in the same field, the chrominance signal recurs out of phase by

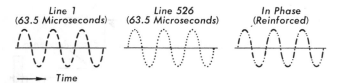

(A) The luminance signal contains harmonics of the line frequency. These harmonics recur in phase during the scanning of the same line in successive frames.

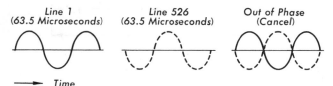

(B) The chrominance signal contains odd harmonics of one-half the line frequency. These harmonics recur out of phase and are effectively canceled.

Fig. 2-12. Brightness variations produced on the screen of a monochrome receiver by the luminance and chrominance signals.

180 degrees. It also recurs out of phase during the scanning of the same line in the succeeding frame. This can be more clearly understood by studying Fig. 2-12B. A cancellation effect occurs as a result of this out-of-phase condition, and the brightness variations produced on the screen of a monochrome receiver by the chrominance signal cannot be perceived by the human eye. Thus, satisfactory monochrome reproduction can be achieved when a composite color signal is being transmitted.

THE VIR SIGNAL

In October of 1975 the FCC reserved line 19 of both fields for insertion of the VIR (Vertical-Interval-Reference) signal. This signal, shown in Fig. 2-13, permits the quality of a color program to be uniformly maintained as it is processed by the studio, network, and transmitter facilities. The VIR signal is added to the video of a color program at a point where the correct blanking level, chrominance reference, luminance reference, and black reference have been established for the composite color signal.

After the VIR signal has been added to a color program, it remains with that program until it is broadcast. This permits adjustments at the various processing facilities to maintain the original color quality of the program. The VIR signal should not be confused with the VIT (Vertical-Interval-Test) signals which have been used for several years as test signals to evaluate the performance of equipment and facilities. The VIT signals are inserted on lines 17 and 18 of each field. (Since the first 21 lines of each field occur during the vertical-blanking interval, neither the VIT nor the VIR signals show up in the picture.)

Although the VIT signals are required for studio-to-transmitter links, the VIR signal is not mandatory at this time. However, it is expected that most tv stations will eventually use the VIR signal. At least one television manufacturer is marketing a receiver with circuitry that utilizes the VIR signal to automatically correct the hue and saturation of the color picture.

SUMMARY

To summarize the material in this chapter, it has been stated that the composite color signal contains a luminance, chrominance, burst, horizontal-sync, and vertical-sync signal. The luminance signal represents the brightness elements of a scene, the chrominance signal represents the colors of a scene, and the burst signal is used as a phase reference.

The use of a specific frequency for the chrominance subcarrier results in an interleaving of the luminance and chrominance signals. This interleaving of signals makes it possible to transmit both the luminance and chrominance signals within the same channel width used for the transmission of a monochrome signal.

The next chapter discusses the formation of luminance and chrominance signals at the transmitter.

QUESTIONS

1. How was the frequency of the subcarrier selected so that interleaving would take place?
2. What is meant by the cancellation effect of the chrominance signal?
3. What happens to the output signal of a doubly balanced modulator when the polarity of the input signal is reversed?
4. What is the nature of the resultant signal produced by the addition of output signals from two doubly balanced modulators when the output signals are of different amplitudes?
5. What purpose does the color burst serve?

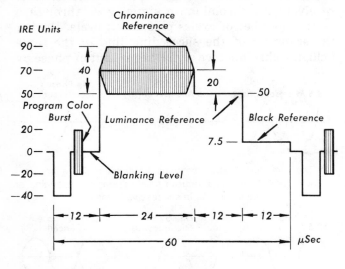

Note: The Chrominance Reference and the Program Color Burst Have the Same Phase.

Fig. 2-13. The VIR signal.

Chapter 3

Makeup of the Color Picture Signal

As has been shown in the foregoing chapter, the color picture signal consists of two separate signals—a luminance signal and a chrominance signal. This chapter discusses the makeup of these two signals at the transmitter.

Shown in Fig. 3-1 is a drawing of the basic components of a tricolor camera, which employs three camera tubes and two dichroic mirrors for the separation of light. Regular image orthicons may be used in this camera; however, newer camera tubes that have a response to light frequencies more nearly like that of the human eye are also used. One of the camera tubes receives only the light frequencies corresponding to the color red and is called the red camera tube. Another tube receives only the light frequencies corresponding to the color blue and is called the blue camera tube. The third tube receives only light frequencies corresponding to the color green and is called the green camera tube.

To illustrate the operation, let us assume that the color camera is focused on a color scene. The light is broken up in the following manner. All the light frequencies pass through the objective lens, which is mounted on the turret, and through a pair of relay lenses. Then the light is affected by the dichroic mirrors. This type of mirror permits all the light frequencies of the spectrum to pass except those of the primary color that it is designed to reflect. By the use of this type of mirror, white light can be separated into the light frequencies of the three primary colors—red, green, and blue.

Through correct placement, only two dichroic mirrors are needed in the color camera. The blue dichroic mirror is positioned at a point indicated by A on the diagram. When light arrives at this point, all the light frequencies except those representing the color blue are passed through the mirror. The frequencies representing the blue portion of the spectrum are reflected. The tilt angle of the mirror at point A is such that the mirror directs the blue light to a front-surface mirror at point C. Then the blue light is reflected onto the face of the blue camera tube.

The light that was passed by the dichroic mirror at point A goes on to the red dichroic mirror positioned at point B. This mirror is designed to pass all the light frequencies except those that represent red. The red light is reflected to a front-surface mirror at point D where it is again reflected so that it falls on the face of the red camera tube.

Both the blue and the red portions of the incoming light frequencies have been removed, and only the green portion remains. This is allowed to fall directly on the face of the green camera tube. In this manner, the light is broken up into the three primary colors.

At the output of the color camera, there are three voltages which are representative of the three colors. These voltages are designated as E_R, E_G, and E_B, where R equals red, G equals green, and B equals blue. From these voltages, the luminance and chrominance signals are formed.

LUMINANCE SIGNAL

The luminance signal is the portion of the color picture signal utilized by monochrome receivers. For this reason, the luminance signal must represent the screen only according to its brightness. It is very similar to the video signal specified for standard monochrome transmission.

The luminosity responses of the eye to the light frequencies of the three primary colors were considered when the specifications for the luminance signal were determined. The responses of the eye

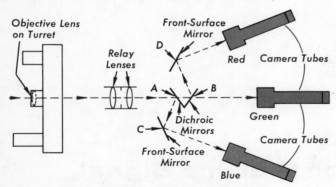

Fig. 3-1. Drawing of the basic components of a tricolor camera.

to these colors can be illustrated through the use of three projectors for colored lights. One projector is for red, one is for blue, and one is for green. If these three projectors are adjusted so that their light outputs are equal, as measured by photoelectric means, white light will be produced when these outputs are superimposed upon each other. When they are separated, the green light will appear to the average observer almost twice as bright as the red and from five to six times as bright as the blue. The red light will appear from

two to three times as bright as the blue light, which will appear the dimmest. This means that the eye is most sensitive to green, less sensitive to red, and least sensitive to blue. This effect was described in Chapter 1.

The specifications for the luminance signal take into consideration the sensitivity of the eye to light frequencies. Definite proportions of each of the color signals from the color camera are used to form the luminance signal. These proportions are: 59 percent of the green signal, 30 percent of the red signal, and 11 percent of the blue signal.

The luminance signal is frequently called the Y signal, and its voltage is designated as E_Y and is expressed as:

$$E_Y = .30E_R + .59E_G + .11E_B \qquad (4)$$

where,
    E_R is the voltage of the red signal,
    E_G is the voltage of the green signal,
    E_B is the voltage of the blue signal.

The drawing in Fig. 3-2 illustrates the manner in which the luminance signal is formed. The scene to be televised consists of a card that has four vertical bars. From left to right, the first bar is

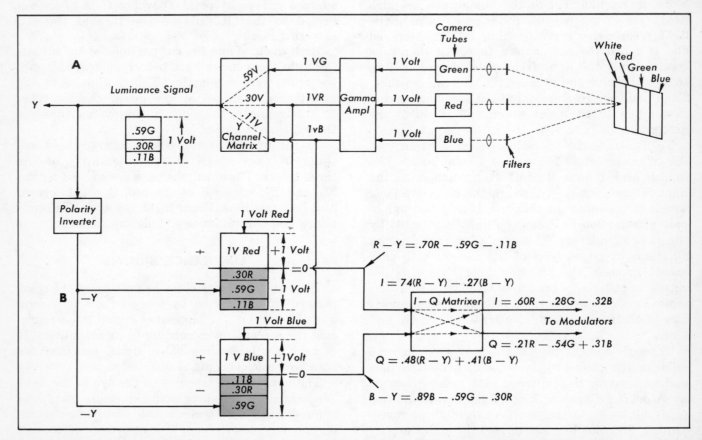

Fig. 3-2. Block diagram showing the units used at the transmitter to form the luminance signal and the color-difference signals.

white, the second is red, the third is green, and the fourth is blue. The color camera is adjusted so that each one of the three color signals from the camera is one volt when the white bar is being scanned. In accordance with equation 4, the luminance signal will also be one volt at this time; consequently, the bright white bar would be produced on the screen of a monochrome receiver tuned to this signal.

When the red bar is being scanned, the blue and green color signals go to zero and the red signal remains at one volt. By equation 4, the luminance signal drops to .30 volt. A gray bar would appear on the screen of the monochrome receiver when this red bar is scanned. The green signal will be one volt, and the blue and red signals will be zero when the green bar is scanned. The luminance signal will have a value of .59 volt and will produce a very light gray bar on the screen of the monochrome receiver. Scanning of the blue bar will cause the blue signal to equal one volt and the voltages of the red and green signals to equal zero. The luminance signal will be .11 volt, and a dark gray bar will appear on the screen of the monochrome receiver.

CHROMINANCE SIGNAL

The chrominance signal must represent only the colors of a scene; therefore, the luminance voltage E_Y is subtracted from each of the three output voltages of the color camera. As shown in Fig. 3-2, this can be done by inverting the polarity of the luminance signal and by combining the resultant signal with each of the three camera signals. This results in three signals which represent red minus luminance, green minus luminance, and blue minus luminance. These signals are denoted as the color-difference signals $E_R - E_Y$, $E_G - E_Y$, and $E_B - E_Y$. The term E_Y in each of these expressions has the value given in equation 4. If the value given in this equation is substituted for E_Y, an expression for each color-difference signal can be obtained in terms of the signals for the three primary colors:

$$E_R - E_Y = .70E_R - .59E_G - .11E_B \qquad (5)$$

$$E_G - E_Y = .41E_G - .30E_R - .11E_B \qquad (6)$$

$$E_B - E_Y = .89E_B - .59E_G - .30E_R \qquad (7)$$

See footnote 1 for derivation of these equations.

The equations for each of the color-difference signals may also be obtained graphically. Refer to Fig. 3-2. Note that when the red bar is scanned, one volt of red signal is applied to the R − Y ma-

trix. The luminance signal at the same time is .30 volt. Since this voltage is applied to the R − Y matrix through a polarity inverter, the value of the minus Y signal is −.30 volt. The combination of voltages fed to the R − Y matrix forms the R − Y signal, and during the scanning time of the red bar, the R − Y signal amplitude is 1.0 minus .30 (or .70 volt). Note that this value conforms with the coefficient of E_R in equation 5.

When the green bar is scanned, the voltage of the red signal is zero, and the luminance signal becomes .59 volt. The output of the R − Y matrix at the same time would be a combination of zero and −.59 (or −.59 volt). Note that this value agrees with the coefficient for E_G shown in equation 5.

When the blue bar is scanned, the voltage of the red signal is zero, and the minus Y signal is −.11 volt. These voltages are combined in the R − Y matrix, and the output voltage is −.11 volt. This is the coefficient E_B in equation 5. The coefficients for the voltages of the color signals in equations 6 and 7 may be obtained in a similar manner.

It is interesting to note that when the white bar is scanned, the voltages of the color-difference signals are equal to zero. It may be recalled from the discussion on divided-carrier modulation in Chapter 2 that, when both of the two modulating signals have zero voltages, the outputs of the balanced modulators also become zero. Thus, during the time that the color camera scans the white portions of a scene, no chrominance signal is developed; and these portions are represented only by the luminance signal.

The same condition is true for any value of gray. Consider that the brightness of the white bar has been reduced by 50 percent. The voltage of each of the three signals at the output of the color camera would equal .5 volt. The voltage of the luminance signal would be the sum of .15, .295, and .055 (or .5 volt); and the voltages of the color-difference signals would still be zero. The bar would be entirely represented by the luminance signal, but its amplitude would be only 50 percent

---

1. As an example of the manner in which equations 5, 6, and 7 are derived, let us consider equation 5. As previously stated, the color-difference signal is formed by subtracting the luminance signal from the color signal. In the case of equation 5, the color signal is E_R. From this we need to subtract E_Y. Equation 4 states that

$$E_Y = .30E_R + .59E_G + .11E_B$$

Therefore,

$$E_R - E_Y = E_R - (.30E_R + .59E_G + .11E_B)$$
$$= E_R - .30E_R - .59E_G - .11E_B$$
$$= .70E_R - .59E_G - .11E_B$$

of the amplitude produced when the camera scans reference white. A monochrome receiver tuned to this signal would reproduce a value of gray that would be halfway between white and black.

It has been stated that only two signals are used to modulate the color subcarrier. These two signals must represent the colors denoted by the three color-difference signals. It was found that a signal equivalent to $E_G - E_Y$ could be reproduced in the color receiver by combining specific proportions of $E_R - E_Y$ and $E_B - E_Y$; therefore, $E_G - E_Y$ is not actually transmitted as such. The combination of $-.51 (E_R - E_Y)$ and $-.19 (E_B - E_Y)$ will produce a signal equivalent to $E_G - E_Y$. (The negative coefficients for these two quantities specify amplitudes of signals having negative polarities.) Thus:

$$E_G - E_Y = -.51 (E_R - E_Y) - .19 (E_B - E_Y)$$

The mathematical proof for this equation is presented in footnote 2.

The two signals used to modulate the color subcarrier are called the I and Q signals. As shown in Fig. 3-2B, these two signals are formed by combining specific proportions of $E_R - E_Y$ and $E_B - E_Y$. This is done because a more faithful reproduction of colors can be obtained.

Shown in Fig. 3-3 is the chromaticity diagram and the NTSC triangle. The axes for the color-difference signals and the I and Q signals can be seen. It should be mentioned that the $R - Y$ and $B - Y$ axes are actually perpendicular to each other. This is also true of the I and Q axes. These facts are not apparent in the two-dimensional drawing in Fig. 3-3. However, if these axes were observed on a three-dimensional facsimile of the Maxwell triangle, they would have a right-angle relationship as dated. Along the axes of the color-difference signals are the colors represented by these signals. Colors from red to bluish-green are

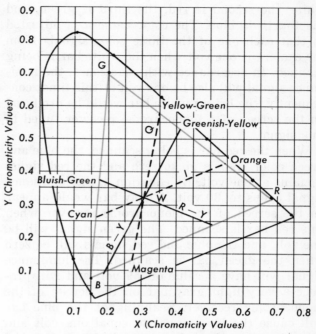

Fig. 3-3. The axes of the color-difference signals and the I and Q signals on the color triangle.

depicted along the $R - Y$ axis and from blue to greenish-yellow along the $B - Y$ axis.

By moving the axes to the I and Q positions, colors from orange to cyan are depicted along the I axis and from magenta to yellow-green along the Q axis. Better reproduction of color is achieved along the I and Q axes than along the $R - Y$ and $B - Y$ axes. This is particularly true in the reproduction of flesh tones, since they lie along the I axis. It was also found that for small areas of color that are well centered in the field of vision, the chromaticity diagram degenerates to a single line. This line is the I axis, and only two fully saturated colors, orange and cyan, are needed to reproduce colors under these conditions. This phenomenon was described in Chapter 1.

The I signal is formed by combining $.74 (E_R - E_Y)$ and $-.27 (E_B - E_Y)$. The Q signal is composed of $.48 (E_R - E_Y)$ and $.41 (E_B - E_Y)$. The equations for I and Q are written as follows:

$$E_I = .74 (E_R - E_Y) - .27 (E_B - E_Y) \qquad (8)$$
$$E_Q = .48 (E_R - E_Y) + .41 (E_B - E_Y) \qquad (9)$$

If the value given in equation 4 is substituted for E_Y in equations 8 and 9, equations for E_I and E_Q can be obtained in terms of the three color signals at the output of the color camera, as follows:

$$E_I = .60E_R - .28E_G - .32E_B \qquad (10)$$
$$E_Q = .21E_R - .52E_G + .31E_B \qquad (11)$$

---

2. From equations 5 and 7 we know that

$E_R - E_Y = .70E_R - .59E_G - .11E_B$,

and that

$E_B - E_Y = .89E_B - .59E_G - .30E_R$.

By substituting these quantities in the equation being proved, we have

$E_G - E_Y = -.51 (.70E_R - .59E_G - .11E_B) - .19 (.89E_B$
$\qquad - .59E_G - .30E_R) = .41E_G - .30E_R - .11E_B$.

The value $.41E_G$ can also be expressed as $E_G - .59E_G$. Therefore,

$E_G - E_Y = E_G - .59E_G - .30E_R - .11E_B$
$\qquad = E_G - (.30E_R + .59E_G + .11E_B)$
$\qquad = E_G - E_Y$.

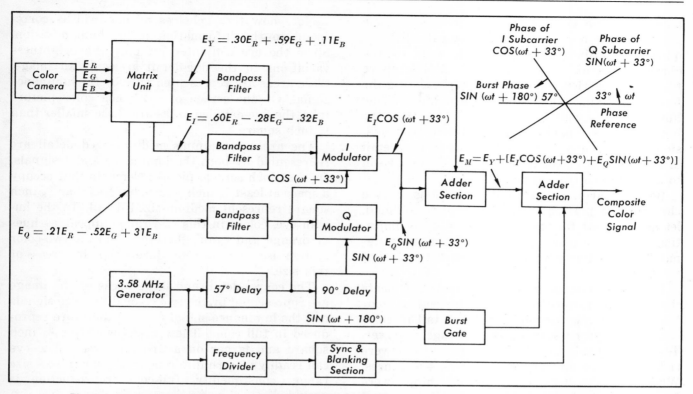

Fig. 3-4. Partial block diagram of a color television transmitter and a diagram showing phase relationships.

The mixing of the three output voltages of the camera to form the luminance signal and the color-difference signals is performed in the matrix unit of the transmitter. See the partial block diagram of a color transmitter shown in Fig. 3-4. The output of the matrix consists of the luminance signal and the I and Q signals. It should be pointed out that signals from the color camera are gamma corrected by passing them through gamma amplifiers. This correction is to compensate for the nonlinear operation of the picture tube. Since gamma correction is also provided in the monochrome transmitter, no block for this is shown in Fig. 3-4.

From the matrix, the luminance signal is fed through a bandpass filter to the adder section. The I and Q signals are fed through bandpass filters to the modulator sections. The I signal modulates a subcarrier, $\cos(\omega t + 33°)$, whereas the Q signal modulates a subcarrier, $\sin(\omega t + 33°)$. In color television, the phase reference ωt is the phase of the color burst plus 180 degrees. The phase angles between the two subcarriers and between them and the color burst are shown in the diagram of Fig. 3-4. Note that the phase of the subcarrier modulated by the I signal is leading the phase of the subcarrier modulated by the Q signal by 90 degrees and is lagging the phase of the color burst by 57 degrees.

These phase relationships must be maintained within a very close tolerance. This is accomplished by using a common source for the 3.58-MHz signals. In addition, the synchronizing signals for horizontal and vertical scanning have the required frequency relationship to the chrominance subcarrier because the frequencies of these signals are derived from the subcarrier generator.

The outputs of the modulators are combined to form the chrominance signal. This signal is fed to an adder section where it is combined with the luminance signal. The output of this adder section is the color picture signal, which is specified by the NTSC standards as follows:

$$E_M = E_Y + [E_Q \sin(\omega t + 33°) + E_I \cos(\omega t + 33°)] \qquad (12)$$

The sync, blanking, and color-burst signals are added to the color picture signal; then the composite color signal is ready for transmission.

BANDWIDTHS OF LUMINANCE AND CHROMINANCE SIGNALS

Let us consider the band limitations of the luminance and chrominance signals. It has been stated before that luminance signal in color transmission must retain the same specifications (within tolerance) that the video signal has in monochrome transmission in order to meet compatibility requirements. Since the upper sidebands of the chrominance subcarrier extend to 4.2 MHz above the picture carrier, the sidebands representing the luminance signal can also extend to 4.2 MHz. This limit is approximately 0.2 MHz greater than the limit of the sidebands in monochrome transmission; consequently, a slight increase in fine detail is available with color transmission in comparison to monochrome transmission.

Before learning the band limitations of the chrominance portion of the color picture signal, let us see what factors led to the specific limitations that were placed upon the chrominance signal. These factors pertain to the characteristics of the human eye.

From the results of colorimetry experiments that were performed prior to the advent of color television, it was known that fine detail in color cannot always be seen by the average observer. Tests were made in which colored objects were reduced in size and viewed at various distances. When this was done, a number of things were found to be true. First, blues become more and more indistinguishable from grays of equivalent brightness as distance increases or size decreases. Second, yellows become indistinguishable from grays. Within the same size range, browns become confused with crimsons and blues with greens, but reds remain clearly distinct from blue-greens. Colors with pronounced blue lose blueness, whereas colors lacking in blue gain blueness. Third, with a further decrease in size, reds merge with grays that have equivalent brightness, blue-greens are indistinguishable from grays. Finally, when viewing extremely small colored objects, the ability to identify color is lost entirely and only a response to brightness remains. (See Chapter 1.)

From the foregoing data, which were obtained from the experiments pertaining to human vision and from tests made with color receivers, the following choices of bandwidth were made.

1. Full-band transmission of the luminance (Y) signal for maximum detail.
2. Moderately wideband, partly single-sideband transmission of a single color-mixture signal (I signal) which represents colors of orange and cyan.
3. Narrow-band, double-sideband transmission of an additional color-mixture signal (Q signal). This signal represents yellow-green and magenta.

Note how these choices of bandwidths correspond to the information about human vision. Since the eye can interpret only the brightness variations in the fine-detail areas of an image, these areas are represented only by the luminance signal. On the screen of a 21-inch color picture tube, the fine-detailed areas are those smaller than ⅛ inch square.

The areas containing medium-sized detail are represented by both the luminance and I signals. On a 21-inch screen, picture elements that occupy a space at least ⅛ inch square but less than ⅜ inch square represent medium-sized detail. To the human eye, colors in this size range appear as hues of orange and cyan; therefore, only a two-color system is used to reproduce color in areas of this size.

The coarse-detail and large areas of the image are represented by the two color-difference signals and the luminance signal and are therefore reproduced in full color. These areas would be ⅜ inch square and larger on a 21-inch screen. The eye can readily discern differences in color in this size range; consequently, full-color reproduction is provided.

Shown in Fig. 3-5 is the passband of the color picture signal. The Q signal is limited to .5 MHz, and both sidebands are transmitted. Both sidebands of the I signal are transmitted for frequencies up to .5 MHz, and only the lower sideband is transmitted for frequencies from .5 to 1.5 MHz.

When frequencies of 0 to .5 MHz are present, Y, I, and Q are being transmitted. A three-color system would therefore be in effect. For frequencies of .5 to 1.5 MHz, only the I signal and the luminance signal are being transmitted. A two-color (orange-cyan) system is in effect at these

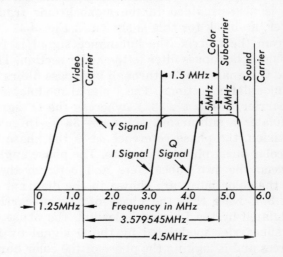

Fig. 3-5. Passband of the color picture signal.

frequencies. For frequencies above 1.5 MHz, only the Y channel is transmitted. The Y signal conveys the fine detail of the picture in terms of brightness variations.

COLOR-BAR CHART

The makeup of the color picture signal can be illustrated by a color-bar chart like the one in Fig. 3-6. Through the use of such a chart, the relative level of each component of the color picture signal can be shown. Let us develop the color picture signal by analyzing the voltages that are present as each bar is being scanned. The colors in the chart are considered to be fully saturated, which means that they are completely free of white light. When these color bars are scanned by the color camera, a color signal is produced. The development of the color signal can be illustrated by showing the waveforms that are formed during the process.

The waveforms that are representative of each of the output signals from the camera are directly below the test bars in column I of Fig. 3-6. The signal waveform shown for the red output of the camera reaches a maximum value when the red bar is scanned. During the scanning of the green and blue bars, this signal drops to zero because no light reaches the red camera tube. The waveform for the green signal reaches maximum during the scanning of the green bar. During scanning of the other two color bars, the green signal drops to zero. Similarly, the blue signal reaches maximum during the scanning of the blue bar and goes to zero during the time the other two bars are being scanned. From the three camera-output signals shown in column I of Fig. 3-6, the signals that make up the color picture signal are formed.

In column II of Fig. 3-6, the waveform for the luminance signal is shown. Equation 4 showed that the luminance signal was formed by taking 30 percent of the red signal, 59 percent of the green signal, and 11 percent of the blue signal and adding them together. These percentages are represented by the waveform of the luminance signal shown. During the scanning of the red bar, E_Y reaches a level of .30 volt; during that of the green bar, it reaches .59 volt; and during that of the blue bar, it reaches .11 volt.

The next signals to be formulated are the three color-difference signals. These are shown in column III of Fig. 3-6. The expressions $E_R - E_Y$, $E_G - E_Y$, and $E_B - E_Y$ for these color-difference signals signify that the voltage value of the luminance signal is subtracted from the voltage values of the camera output signals. Subtraction of the instantaneous value of the luminance signal from the instantaneous value of E_R will give the instantaneous voltage value for $E_R - E_Y$. Subtraction of .30 volt from 1.00 volt leaves a value of .70 volt for the $E_R - E_Y$ signal during the scanning of the red bar. During the scanning of the green and blue bars, $E_R - E_Y$ will be negative because there is no voltage from the red output in the camera. It will be −.59 volt during the scanning of the green bar and −.11 volt during the scanning of the blue bar. The same method is followed for obtaining the values of $E_G - E_Y$ and $E_B - E_Y$ as they are shown in column III of Fig. 3-6. Note that the values

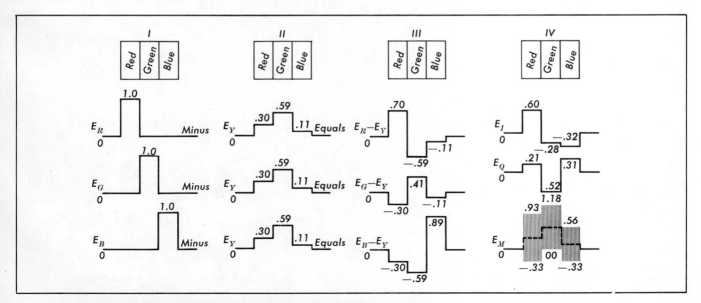

Fig. 3-6. Waveforms produced when a color-bar chart consisting of red, green, and blue is being scanned.

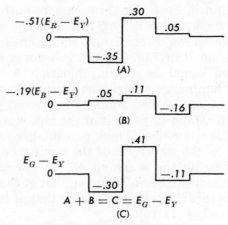

Fig. 3-7. Formation of $E_G - E_Y$.

which appear on the waveforms are the same as those which appear in equations 5, 6, and 7.

It has been stated that the color-difference signal $E_G - E_Y$ is not transmitted. It is obtained in the receiver by combining $-.51$ $(E_R - E_Y)$ and $-.19$ $(E_B - E_Y)$. This has been proved mathematically. The manner in which $E_G - E_Y$ is obtained can be shown graphically by using the waveforms of the color-difference signals. If the numerical values of the $E_R - E_Y$ waveform shown in Fig. 3-6 are multiplied by the factor of $-.51$, waveform A of Fig. 3-7 can be obtained. Waveform B in the same figure is the result of multiplying the numerical values of the waveform for $E_B - E_Y$ in Fig. 3-6 by the factor of $-.19$. The addition of waveform A and waveform B in Fig. 3-7 results in waveform C. This is the same waveform that is shown in Fig. 3-6 for $E_G - E_Y$. Again, it has been shown that the color-difference signal representing green can be recovered by proportionately mixing the other two color-difference signals.

In column IV of Fig. 3-6, waveforms for the I and Q signals are presented. The I waveform is obtained by adding .74 of the $E_B - E_Y$ signal and $-.27$ of the $E_B - E_Y$ signal. The Q waveform is obtained in a similar manner; however, this signal is formed by combining .48 of the $E_R - E_Y$ signal and .41 of the $E_B - E_Y$ signal. The numerical values shown in Fig. 3-6 for the I and Q signals correspond to those previously given in equations 10 and 11.

The last waveform in Fig. 3-6 represents the color picture signal E_M, which is the signal that is transmitted. The numerical values shown with the E_M waveform specify the levels of the maximum excursions of the chrominance signal. To determine these levels, the values of I and Q are added together vectorially, and then the results are added

to and subtracted from the luminance levels. Fig. 3-8 shows how the red portion of the signal is formed by this vectorial method. The resultant vector, which represents the red portion, is found by marking off the values of I and Q on their respective vectors. As shown on the waveforms for I and Q, the I value for red is .60 and the Q value for red is .21. The resultant vector is drawn from the origin of the vectors to the opposite corner of the parallelogram. The magnitude of the resultant vector is found by taking the square root of the sum of the squares. In algebraic form this process can be written:

$$\sqrt{I^2 + Q^2}$$

Then substituting the values for I and Q and solving we have:

$$\sqrt{(.60)^2 + (.21)^2} = .63$$

The same procedure has been followed for determining the magnitude of the chrominance signal when it represents fully saturated colors of green and blue. When a fully saturated green is transmitted, the chrominance signal has a relative amplitude of .59; when a fully saturated blue is transmitted, the chrominance signal has a relative amplitude of .45.

Let us now consider the waveforms produced during the scanning of a color-bar chart that not only includes the three primary colors but also their complementary colors. Such a color-bar chart and the associated waveforms are shown in Fig. 3-9. The colors represented are green, yellow, red, magenta, blue, and cyan. These colors are assumed to be fully saturated.

The waveforms of first concern are those at the output of the camera. During scanning of the

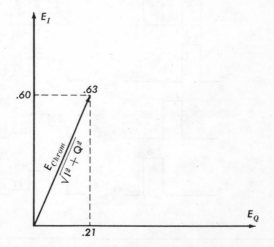

Fig. 3-8. Vectorial addition of I and Q amplitudes.

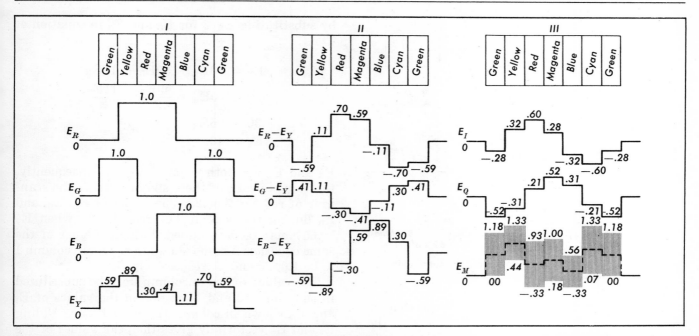

Fig. 3-9. Waveforms produced when a color-bar chart consisting of three primary colors and their complements is being scanned.

green bar, a signal is produced at the output of the green camera tube only. When scanning yellow, the camera has signal outputs from both the red and green camera tubes. This is to be expected because yellow contains both red and green light.

While the red bar is being scanned, a signal is present at the output of the red camera tube only. Magenta is a combination of red and blue; therefore, there will be outputs from both the red and blue camera tubes while the magenta bar is being scanned.

During the scanning of the third primary color, blue, there is a signal at the output of the blue camera tube only. There is an output at both the green and blue camera tubes while the cyan bar is being scanned. These outputs are produced because cyan is a mixture of green and blue.

The next waveform shown in Fig. 3-9 is the luminance signal. The levels of the luminance signal for red, green, and blue are the same as those shown in Fig. 3-6. Yellow, however, produces a luminance value of .89, which is made up of 59 percent of green plus 30 percent of red. Magenta has a luminance level of .41, or 30 percent of red plus 11 percent of blue. Cyan contains blue and green; therefore, the luminance level for cyan consists of 11 percent of blue plus 59 percent of green and has a value of .70. Yellow has the highest luminance level.

The color-difference waveforms are formed in the same manner as before by subtracting the luminance signal from each of the three color sig-

nals from the camera. If this is done with the waveforms shown in Fig. 3-9, the resultant waveforms will be as shown in column II for $E_R - E_Y$, $E_G - E_Y$, and $E_B - E_Y$.

The waveforms for I and Q are obtained by proportionately mixing the color-difference signals $E_R - E_Y$ and $E_B - E_Y$. By taking .74 $(E_R - E_Y)$ and adding the result to $-.27$ $(E_B - E_Y)$, the waveform for the I signal shown in column III of Fig. 3-9 is obtained. The waveform for the Q signal is formed by taking .48 $(E_R - E_Y)$ and adding this to .41 $(E_B - E_Y)$.

The chrominance signal combined with the luminance signal is shown as the last waveform of Fig. 3-9. The chrominance values of this waveform were determined by vector addition of the I and Q signals, as previously described, and the resultant values have been added to and subtracted from the luminance values.

The relative saturation of a color is conveyed by the ratio between the amplitudes of the chrominance and the luminance signals. The more highly saturated the color, the higher the ratio becomes. Moreover, the ratio remains fixed for a color with a given saturation regardless of the brightness of the color.

Shown in Fig. 3-10 is a color-bar chart together with the various signals that are produced at the color transmitter as the camera scans the pattern. The pattern consists of three bars which are all red and which have specific saturation and brightness levels. Red No. 1 is a fully saturated red with

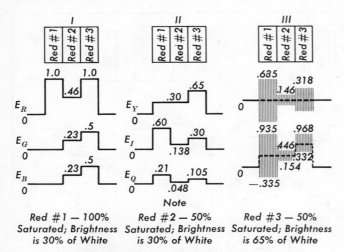

Note

Red #1 — 100%
Saturated; Brightness
is 30% of White

Red #2 — 50%
Saturated; Brightness
is 30% of White

Red #3 — 50%
Saturated; Brightness
is 65% of White

**Fig. 3-10. Signals produced under various conditions
of saturation and brightness.**

a brightness that is equivalent to 30 percent of white. Red No. 2 has the same brightness as the bar on the left, but it has a saturation of only 50 percent. Red No. 3 is a red with a saturation of 50 percent, but it has a brightness level of 65 percent of white.

In column II of this figure, the value of the brightness or luminance signal E_Y for each bar is shown to be as specified. Waveforms of the three signal outputs of the color camera are pictured in column I. During the scanning of fully saturated red No. 1, the amplitude of the red signal is unity and the blue and green signals are at zero. For each of the bars with a saturation of 50 percent, the green and blue signals are shown to have equal amplitudes and the red signal is shown to have twice the amplitude of either of the other two.

Let us see why these amplitudes have this relationship. Any color that is less than fully saturated contains white light. Since white light is produced by a combination of equal values of the three primaries, a desaturated color may be defined as a mixture of a pure color and its complementary color. A color that has a saturation of 50 percent is produced when one half of the total light is contributed by the primary color and the other half is contributed by the complementary color. Since the complement of red is produced when blue and green are combined in equal amounts, it can be established that for a red having a saturation of 50 percent:

$$\frac{E_R}{2} = E_G = E_B$$

The brightness value of red No. 2 in Fig. 3-10 is .30; therefore, the value of E_R can be determined

by substituting $E_R/2$ for E_G and E_B in equation 4. Thus:

$$E_Y = .30E_R + .59E_G + .11E_B$$
$$.30 = .30E_R + \frac{.59E_R}{2} + \frac{.11E_R}{2}$$
$$.30 = .65E_R$$
$$E_R = .46$$

E_G and E_B are both equal to $E_R/2$; consequently, the value of each of these signals during the scanning of red No. 2 is .23. The values of E_R, E_G, and E_B during the scanning of red No. 3 (when E_Y = .65) may be determined through the use of the same equation. These values are shown in column I to be 1.0, .5, and .5, respectively.

The values shown in column I can be substituted in equations 10 and 11 to obtain the values of E_I and E_Q shown in column II for each bar. Adding E_I and E_Q vectorially gives the peak values of the color subcarrier, and then the color picture signal E_M is formed when the color subcarrier is superimposed on the luminance signal E_Y. See column III.

The chrominance-signal amplitude during the scanning of a color is not necessarily a measure of the saturation of that color. Instead, the saturation is dependent upon the ratio of the chrominance amplitude to the luminance amplitude. In going from red No. 3 to red No. 1, for example, the chrominance amplitude increases and the luminance amplitude decreases. This definitely indicates an increase in saturation. (The ratio of chrominance to luminance increases.) In going from red No. 2 to red No. 3, however, the chrominance and luminance amplitudes both increase. Without considering the ratio of the chrominance amplitude to the luminance amplitude, the observer cannot assume that the saturation has increased. Actually, these two colors have the same saturation because this condition has been established.

In the foregoing example, only the color red has been considered. The ratio of chrominance to luminance for a fully saturated red is:

$$\frac{.635}{.3} = \frac{2.1}{1}$$

As the hue varies, the ratio will also vary. Table 3-1 is a list of the chrominance and luminance values and the ratios for the three primary colors and their complements under fully saturated conditions.

Table 3-1. Chrominance-to-Luminance Ratios for Fully Saturated Colors

Color	Chrominance Value	Luminance Value	Ratio
Red	.63	.30	2.1 to 1
Yellow	.45	.89	.5 to 1
Green	.59	.59	1.0 to 1
Cyan	.63	.70	.9 to 1
Blue	.45	.11	4.05 to 1
Magenta	.59	.41	1.44 to 1

VECTOR RELATIONSHIP OF COLOR SIGNALS

A change in hue causes a corresponding change in the phase of the chrominance signal. This phase equals the phase difference between the chrominance signal and the color burst. For example, if it is stated that the chrominance signal has a phase angle of 57 degrees, this means that the phase difference between the chrominance signal and the reference signal is 57 degrees.

Phase relationships are most conveniently shown by the use of vectors. Fig. 3-11 contains vector diagrams showing the phase displacement of chrominance signals as they are related to the reference burst. It can be seen that each signal is associated with a particular phase angle. A change in phase of the chrominance signal is the only way a change in hue can be conveyed.

Each vector shown in Fig. 3-11 specifies a position of the chrominance signal at a certain instant during the scanning of a scene. Actually, the chrominance vector and the reference vector are constantly rotating during the time that color is being transmitted. The chrominance vector changes in phase whenever there is a change in hue, and it changes in length in accordance with changes in brightness or saturation.

The functions of the chrominance vector and reference vector can be compared to the counter-clockwise rotations of two wheels, each having a radial line drawn on it. One wheel is rotating at a constant speed or frequency, and the second is rotating at a variable speed. Since the first wheel rotates at a constant speed, let the line on it serve as a reference to which the rotation of the second wheel can be compared. The operation of this constant-speed wheel with its line of radius is similar to that of the reference-burst vector which rotates at a constant frequency of 3.58 MHz. The speed of the second wheel is made to change, either to increase or to decrease. It can run at speeds higher than, lower than, or equal to the speed of the reference or constant-speed wheel. The line on the

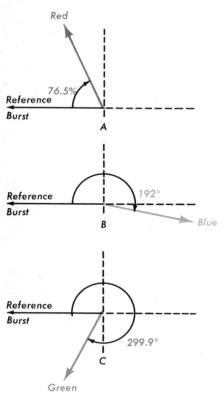

Fig. 3-11. Vector diagrams showing the phase differences between the chrominance signal and the burst signal during the transmission of the three primary colors.

wheel of variable speeds is analogous to that of the chrominance vector. If the two wheels are brought to a sudden stop, the phase difference between the positions of the two lines on the wheels can be determined.

This is in effect what has been done with the vectors that are shown in the drawings of Fig. 3-11. The rotating chrominance vector and reference-burst vector have been halted at specific times. Part A of Fig. 3-11 shows the positions of the vectors at one specific time. The position of the chrominance vector at this time is lagging the position of the reference burst by 76.5 degrees. This difference in phase is representative of a particular color. The chrominance vector at this phase angle represents a red hue. Parts B and C of Fig. 3-11 show the phase differences between the chrominance vector and the vector of the reference burst when blue and green are transmitted. In part B, the chrominance vector is lagging the vector of the reference burst by 192 degrees, and this represents a blue hue. In part C, it is lagging the reference burst by 299.9 degrees. With the chrominance vector in this position, a green hue is being represented.

Let us say that a color-bar chart such as the one that was shown in Fig. 3-6 is being scanned. The vectors will have the following phase relationships. While the red bar is being scanned, the phase relationship between the chrominance signal and the burst signal will be shown in Fig. 3-11, part A. The chrominance vector will lag the vector of the burst signal by 76.5 degrees. This relationship will remain the same during the entire scanning of the red bar. When the camera starts to scan the green bar, the chrominance vector will change position to correspond to that shown in part C of Fig. 3-11. It will remain at this lagging angle of 299.9 degrees until the green bar is entirely scanned. Its position will change to correspond to that shown in part B of Fig. 3-11 when the blue bar is scanned. The vectors for the chrominance and reference signals keep spinning at a constant frequency, and the phase difference of 192 degrees remains the same while the color blue is scanned.

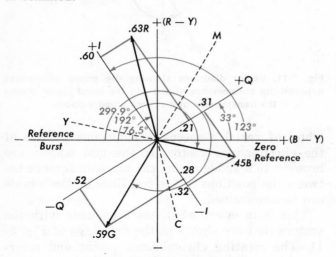

Fig. 3-12. The color-phase diagram.

Fig. 3-12 shows a color-phase diagram representing the color vectors of the system in composite form. It shows the positions of the vectors for the three primary colors employed in color television, their three complementary colors, the I and Q signals, and the color-difference signals. The positions of the vectors for the primary colors are as follows: (1) for red, reference burst minus 76.5 degrees; (2) for blue, reference burst minus 192 degrees; and (3) for green, reference burst minus 299.9 degrees. This means that when one of these colors is being transmitted, the phase angle of the chrominance signal is lagging the phase angle of the burst signal by the number of degrees designated.

The vectors for the three complementary colors are shown by dashed lines in opposite directions from those for the three primary colors. The vector for each complementary color is 180 degrees out of phase with the vector that represents the corresponding primary color. Yellow is represented by a chrominance signal which has a phase angle lagging that of the burst signal by 12 degrees. A chrominance signal representing magenta has a lagging phase angle of 119.9 degrees with respect to the phase angle of the reference burst. Cyan has a lagging phase angle of 256.5 degrees with respect to the phase angle of the reference burst.

The I and Q vectors are shown in relationship to the zero reference, which has a phase of 180 degrees with respect to the reference burst. The Q signal leads the zero reference by 33 degrees. (This angle can also be expressed as an angle that lags the reference burst by 147 degrees.) The I vector leads the zero reference by 123 degrees. (This angle can also be expressed as an angle which lags that of the reference burst by 57 degrees.) The phase angle between the I signal and the R − Y signal and between the Q signal and the B − Y signal is 33 degrees.

SUMMARY

The foregoing discussion has covered the make-up of the color picture signal. It has shown how the individual signals are formed and what each signal consists of. The mixing proportions of the signals that combine to form the color picture signal have been discussed. It has also been shown how brightness, hue, and saturation are conveyed by this signal. The makeup of the color picture signal is of great importance because the color receiver must use this signal to reproduce the original colors of the televised scene.

Fig. 3-13 graphically illustrates the major points that have been covered. It shows that three different signals combine to form the color picture signal. As shown in the drawing by the arrows progressing downward, the two signals E_I and E_Q combine to form the chrominance portion of the color picture signal. The arrows pointing upward show the luminance level of the color picture signal. The chrominance signal is superimposed on the luminance signal which determines the average modulation levels of the chrominance signal. The relative saturation of a color is determined by the ratio between the amplitudes of the chrominance and luminance signals. Hue is determined

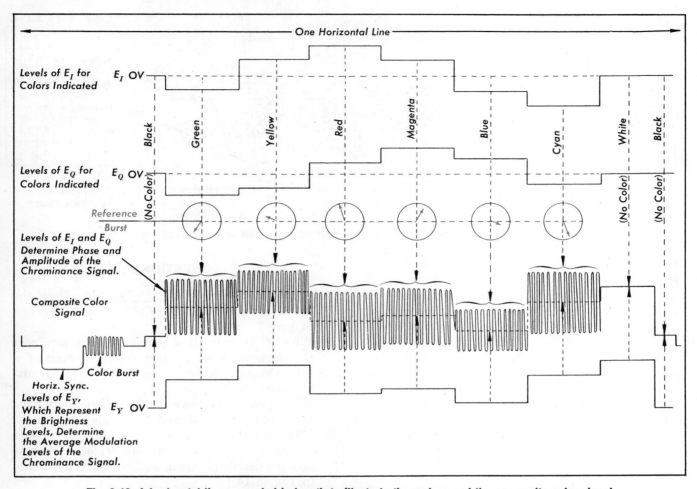

Fig. 3-13. A horizontal line expanded in length to illustrate the make-up of the composite color signal.

by the phase of the chrominance signal with respect to the reference burst. The vectors shown in the circles represent the phase for each color. With blanking, sync, and color burst added to the color picture signal, the composite color signal is formed.

A good way to remember the signal polarities associated with various colors is through the use of the color triangle. During this discussion, the reader may have noted that an I, Q, or color-difference signal may have a negative polarity for some colors and for other colors any one of these signals may have a positive polarity. By studying the color triangles of Fig. 3-14, the reader may find it easier to remember which colors produce an I, Q, or color-difference signal that is negative and which colors provide a positive signal.

On the color triangle of Fig. 3-14A, the polarity of the I signal for each color is given. The colors that fall to the right of the Q axis are represented by a positive I signal, and the colors to the left of the Q axis are represented by a negative I signal.

For instance, when blue, cyan, or green is transmitted, the polarity of the I signal is negative. When magenta, red, or yellow is transmitted, the I signal is positive in polarity. Fig. 3-14B shows the polarity of the Q signal for each color. As can be seen, the colors that lie above the I axis are represented by a negative Q signal, and those lying below the axis produce a positive Q signal. The polarity of the Q signal is negative when cyan, green, or yellow is transmitted, and it is positive when blue, magenta, or red is transmitted.

A composite drawing of triangles A and B is shown in Fig. 3-14C. Notice that the I and Q signals representing colors in the upper left-hand section of the triangle are negative and that the signals representing colors in the lower right-hand section are positive. Colors lying in the other two sections produce I and Q signals which are opposite in polarity. For instance, the Q signal is positive for blue, but the I signal is negative.

The key for determining the correct polarity for each of these signals is in knowing the location of

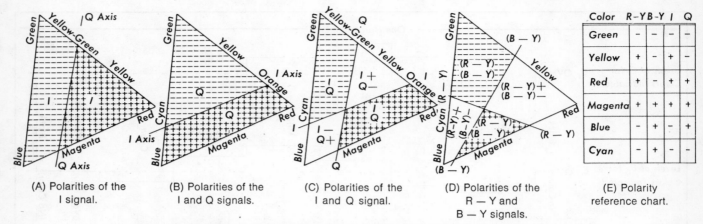

(A) Polarities of the I signal.

(B) Polarities of the I and Q signals.

(C) Polarities of the I and Q signal.

(D) Polarities of the R − Y and B − Y signals.

(E) Polarity reference chart.

Color	R−Y	B−Y	I	Q
Green	−	−	−	−
Yellow	+	−	+	−
Red	+	−	+	+
Magenta	+	+	+	+
Blue	−	+	−	+
Cyan	−	+	−	−

Fig. 3-14. Color triangles showing signal polarities associated with various colors.

the colors on the triangle and in remembering the negative and positive areas shown in Figs. 3-14A and 3-14B. With this knowledge, the polarity of each signal for any color can be easily determined.

Fig. 3-14D shows the polarities of the R − Y and B − Y signals for each color on the color triangle. This drawing can be used to determine the polarities of the R − Y and B − Y signals in the same manner that Fig. 3-14C can be used to determine the polarities of the I and Q signals. Fig. 3-14E lists the various signal polarities in tabular form for those who may prefer this type of presentation.

QUESTIONS

1. What can be said about the outputs from the tubes in a color camera if the scene is colorless or gray?

2. What fully saturated primary color produces the darkest shade of gray on the screen of a monochrome receiver tuned to a color transmission?

3. What are the bandwidth limitations that have been set on the I, Q, and Y signals?

4. What are the polarities of the I and Q signals for each of the three primary colors? (See if you can determine them without having to refer to Fig. 3-14.)

5. What three signals are used to make up the color picture signal?

Section II

Color Receiver Circuits

Chapter 4

RF and I-F Circuits

The study of color receiver circuits has been divided into two categories. The first deals with the color receiver circuits that are similar to those used in monochrome receivers. The material presented deals mainly with the changes that have been made in these sections in order to achieve color reception. If a particular color receiver circuit is very similar to its counterpart in the monochrome receiver, very little discussion is given to it since the reader should already have an understanding of its operation from his past experience in black-and-white television.

The second category has to do with color receiver circuits that are specifically designed for handling the chrominance portion of the composite color signal. Since these circuits are not used in monochrome receivers, a detailed discussion of them will be presented. The following discussions resulted from investigation of color receivers that were made by several manufacturers. The circuits were selected from individual receivers, and they are intended to serve as representative circuits which perform a given function. In some cases, two or more circuits that accomplish the same function are shown, but this is done only when there is a difference in the design or method of operation. Of course, space does not permit a discussion of each of the many circuit variations now in use.

Rapid developments in color television are being made; however, these developments will not change the basic function of the circuits in a color receiver. Any knowledge obtained from this study will be a definite aid in understanding color television and the color receiver circuits of the future.

A complete block diagram of a color receiver is shown in Fig. 4-1. The shaded blocks represent sections that are similar to those used in monochrome receivers. The unshaded blocks represent the sections used only in color receivers. This block diagram is referred to throughout the chapters covering color receiver circuits.

The sections to be discussed in this chapter are the tuner, the video i-f, and the sound i-f sections. From the block diagram in Fig. 4-1, it can be seen that these sections of a color receiver are not too much different from those of a monochrome receiver. With the exception of the block representing the sound detector, this drawing could represent the rf, i-f, and sound circuits of any television receiver; nevertheless, the following discussions about circuits show some important differences.

RF TUNER

The function of the rf tuner in a color receiver is the same as in monochrome receivers, and the physical appearance of the unit has not changed. However, the rf circuits in a color receiver must have one feature that is not necessarily required of the rf circuits in monochrome receivers. This concerns the allowable tolerance in the frequency response of the tuner.

It has been previously pointed out that a color picture signal comprises both a luminance and a chrominance signal. As seen in Fig. 4-2, the bandwidth of this signal extends from 0.75 MHz below to 4.2 MHz above the picture carrier and falls to zero at 1.25 MHz below and at slightly less than 4.5 MHz above. The bandwidth of tuners used in early monochrome receivers often extended to only about 3 MHz. If a tuner designed for an early monochrome receiver were to be used in a color receiver, nonuniform amplification of frequencies might result and cause poor color reception. Although a tilt or sag in the response of the rf tuner might be compensated for in the i-f amplifier section, it is necessary to provide uniform

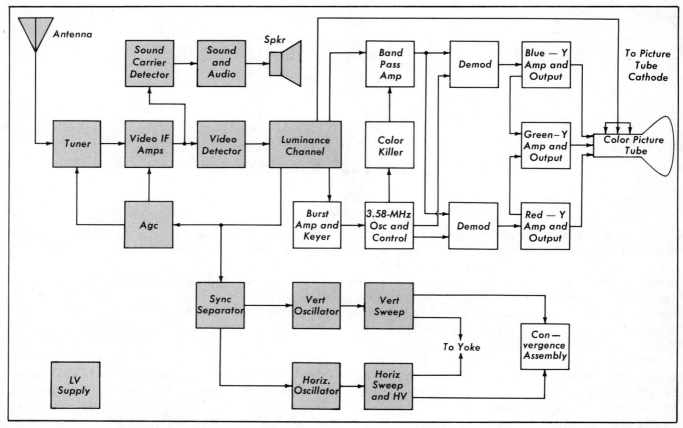

Fig. 4-1. Complete block diagram of a color television receiver. Sections similar to those used in a monochrome receiver are shown shaded. Sections used only in color television receivers are unshaded.

bandpass characteristics in the tuner for proper reception of color telecasts on all channels. An rf circuit that has a frequency response similar to that shown in Fig. 4-3 would produce excellent results in a color receiver.

The vhf tuner used in a typical tube-type color television receiver is shown in Fig. 4-4. The varactor diode (X301) in the oscillator circuit is utilized by the aft (automatic fine tuning) circuit to adjust the oscillator for optimum tuning. The aft voltage applied to the varactor diode determines its capacitance and, therefore, the operating frequency of the oscillator. The uhf tuner used in the

same television receiver is shown in Fig. 4-5. The uhf tuner must have the same broad-band characteristics as the vhf tuner in order for the receiver to produce a good color picture on uhf channels. Note that varactor diode X302 is the aft control diode for the uhf tuner.

Many television tuners are now designed to be used in both monochrome and color receivers. A good example of such a design is the RCA tuner Model KRK140A, B, shown in Fig. 4-6. This tuner employs a 6DS4 Nuvistor as the rf amplifier, along with transistors operating as the oscillator and mixer stages. This circuit configuration provides a high signal-to-noise ratio and contributes greatly

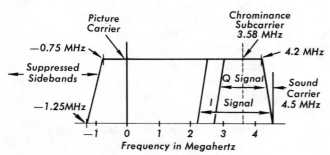

Fig. 4-2. Frequency distribution of the color picture signal.

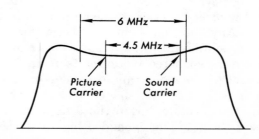

Fig. 4-3. Ideal frequency response of a color television tuner.

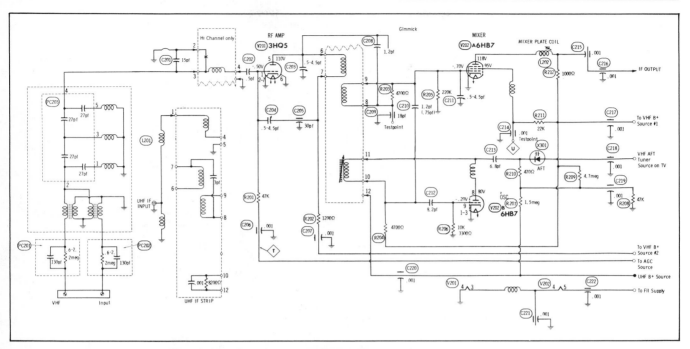

Fig. 4-4. Vhf tuner used in a typical tube-type color television receiver.

toward improved reception in both local and fringe areas.

Some late-model color television receivers are equipped with varactor tuners. The schematic diagram of a typical varactor tuner is shown in Fig. 4-7. The unique feature of this type of tuner is the complete elimination of mechanical tuning devices within the tuner. Since there are no switch con-

tacts in the tuner, its dependability is greatly increased. Tuning is accomplished in the varactor tuner by utilizing the characteristics of varactor diodes that cause them to behave as capacitors when they are reverse-biased. The amount of reverse-bias voltage applied to the varactor will determine its capacitance. As shown in Fig. 4-7, varactor diodes are used for tuning the antenna-

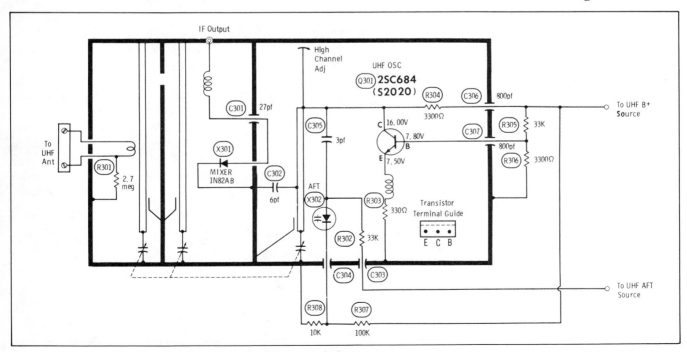

Fig. 4-5. Uhf tuner used with the vhf tuner shown in Fig. 4-4.

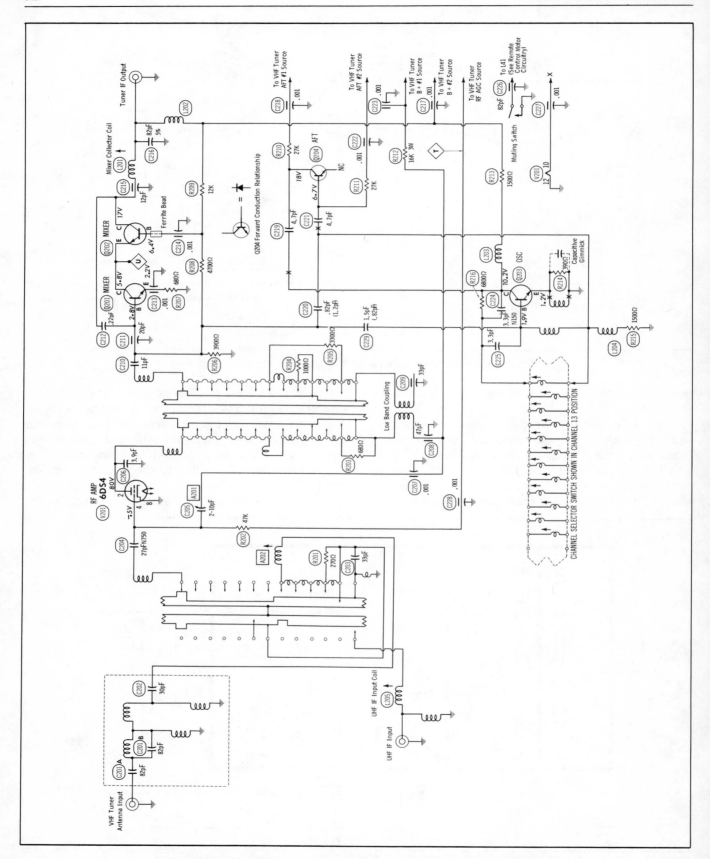

Fig. 4-6. RCA KRK140A tuner used in monochrome or color television receivers.

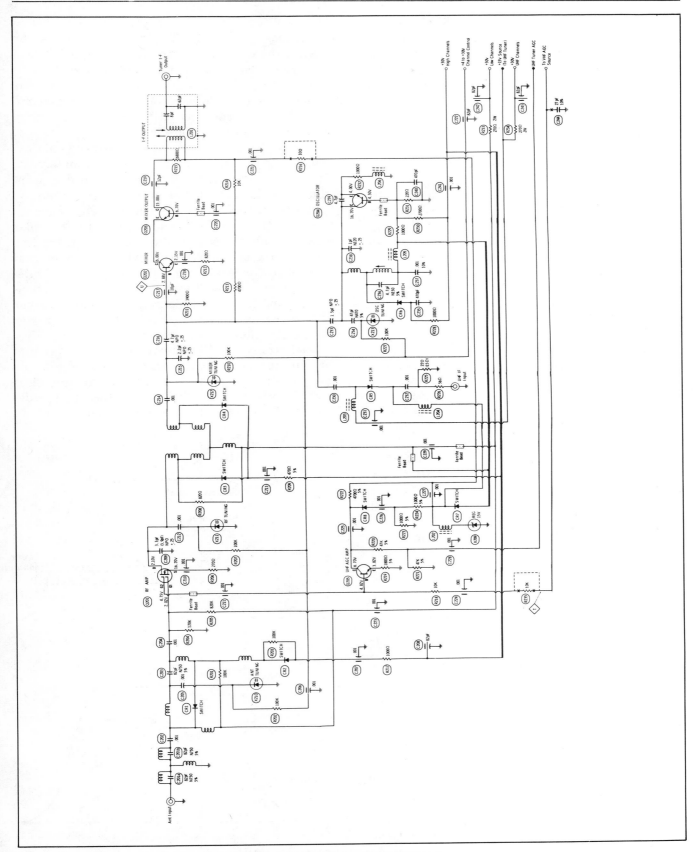

Fig. 4-7. Schematic diagram of a varactor tuner.

input, rf-amplifier, mixer, and oscillator circuits. The desired channel is selected by varying the common control voltage applied to each of the varactor diodes.

Channel selection is accomplished by use of a rotary switch or push-button switches to apply the proper voltage to the varactors for a particular channel. The voltage supplied to the varactors for a given switch position or push button is determined by a separate potentiometer for each channel. Diodes CR1, CR2, CR3, CR4, CR5, and CR6 are bandswitching diodes that are used to switch between the two vhf bands. When a positive voltage is applied to the anode of the diodes, they are forward-biased to switch the tuner to the high-band position.

The improvement in the manufacture of solid-state devices is the primary reason behind the greatly improved performance of tuners. More stringent control of bandpass characteristics has developed, even as circuit design is simplified.

Past experience has shown that rf tuners in monochrome receivers are quite stable and that they present only a moderate number of servicing problems. Tuners in color receivers present additional servicing problems because they must operate more precisely than those in monochrome receivers. A defective tuner that attenuates high frequencies might continue to provide satisfactory results during monochrome transmission; however, such a tuner would probably cause poor reception during color transmission. Such a condition would obviously result in a complaint. When the tuner in a color receiver is being serviced, particularly during alignment, the bandpass requirement of the tuner must be kept in mind. Many of the compromises that are common practice in servicing tuners for monochrome receivers cannot be made in tuners for color receivers.

VIDEO I-F AMPLIFIERS AND VIDEO DETECTOR

Although the bandpass characteristics of the video i-f section are more restrictive in a color receiver, their function is essentially the same as in monochrome receivers. Before examining any circuits, let us consider what is required of the video i-f section.

First of all, the purpose of the section is to provide amplification and selectivity to a specific band of frequencies, such as those illustrated in Fig. 4-8A. Note that the frequency of the sound carrier is below that of the picture carrier. The relative positions of the frequency components have been

reversed from the relative positions occupied in the transmitted signal. This reversal is caused by having the oscillator in the tuner operating at a frequency above that of the received signal.

The upper and lower sidebands receive their names from the original transmitted rf signal and should not be confused with the positions they occupy on the i-f response pattern.

The i-f bandpass of a color receiver should extend from .75 MHz above to approximately 4.2 MHz below the i-f picture carrier in order to include both luminance and chrominance signals.

A curve illustrative of good overall frequency response through the rf and i-f sections of a color receiver can be seen in Fig. 4-8A. Compare this curve to the one in Fig. 4-8B which is representative of the overall frequency response of late-model monochrome receivers. It can be particularly noted that the response of the color receiver is very critical in the region of the sound carrier where the slope of the curve is very steep. Frequencies only .35 MHz away from the maximum attenuation point at the sound-carrier frequency are provided with at least 90-percent amplification. The reason for this is that the upper sidebands of the chrominance subcarrier extend to

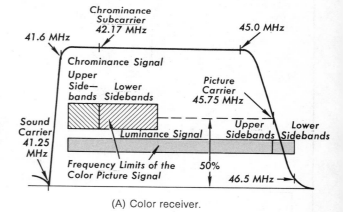

(A) Color receiver.

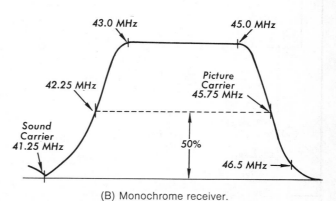

(B) Monochrome receiver.

Fig. 4-8. Rf and i-f response curves.

this portion of the frequency curve. To illustrate this, the frequency limits of the color picture signal have been superimposed on the response curve of Fig. 4-8A. Although the frequency response indicated by the curve in Fig. 4-8B would produce good results during a monochrome transmission, it would severely attenuate the chrominance signal during color transmission. This loss of chrominance would result in poor color reproduction or a complete loss of color.

The video detector demodulates the i-f signal so that the luminance, chrominance, and sync signals are available at the output of the detector circuit. A crystal diode with an i-f filter is commonly used for this purpose. The video detector in a color receiver usually employs a sound-carrier trap in its input. This trap attenuates the sound carrier and ensures against the development of an undesirable 920-kHz beat frequency (which is the frequency difference between the sound carrier and the color subcarrier). When the sound carrier is attenuated in this manner, the sound takeoff point is located ahead of the video detector.

With these requirements in mind, let us examine the video i-f and video-detector circuits in a recent color receiver. Fig. 4-9 shows a schematic diagram of these circuit configurations employed in RCA Chassis CTC38A. As was the case in the rf section, the evolution of solid-state devices has greatly simplified circuit design. In contrast to early-model color receivers where use of five stages of video i-f amplification was not uncommon (to attain proper bandwidth and signal level), hybrid color receivers of the present era require only three stages.

The video i-f output signal developed by the tuner is applied to the first video i-f amplifier through a double-tuned link-coupling circuit. This tuned link coupling consists of the mixer coil (part of tuner), variable capacitor C13 (3–15 pF), and the i-f input coil, L2. Separate 41.25-MHz and 47.25-MHz traps are connected in the input circuit; adjacent sound rejection is accomplished through the use of a 12,000-ohm resistor (R269) in conjunction with the 47.25-MHz trap. Output of the video i-f or link-coupling circuit is applied

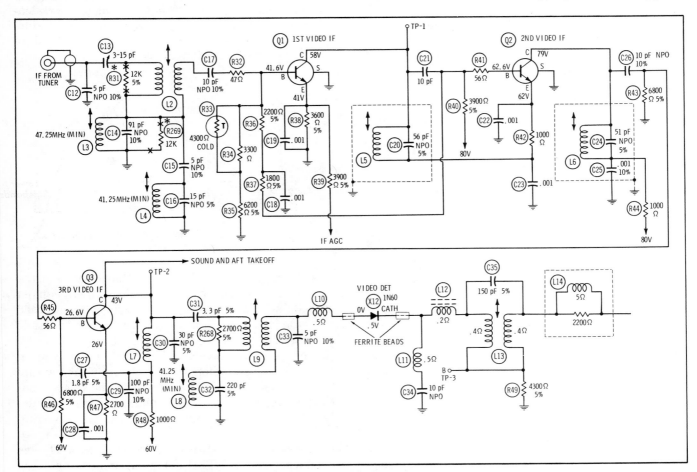

Fig. 4-9. Solid-state video i-f and detector circuits used in an RCA CTC38A color receiver.

to the base of the first video i-f amplifier transistor via the 10-pF capacitor (C17) and the 47-ohm resistor (R32). The signal present at the collector of Q1 is an amplified reproduction of the input signal. This signal is then applied to the base of the second video i-f amplifier through a coupling network comprising a 10-pF capacitor (C21) and a 56-ohm resistor (R41). This type of coupling ensures a good impedance match between the collector of the first video i-f transistor and the base of the second video i-f transistor.

The first and second video i-f stages in the schematic shown in Fig. 4-9 are stagger-tuned. The first video i-f transformer (L5) peaks near 42.17 MHz. The output signal present at the collector of the second video i-f transistor is impedance-coupled to the base of the third video i-f transistor. The third video i-f stage is arranged in a conventional grounded-emitter circuit configuration and includes a double-tuned transformer in the collector circuit. In the configuration illustrated in Fig. 4-9, the double-tuned circuit employs two separate coils. The primary consists of coil L7 connected in the collector circuit of Q3; the secondary is made up of coil L9. The primary-to-secondary coupling is provided by a 3.3-pF capacitor (C31). The value for capacitor C31 is selected to provide maximum efficiency of coupling from primary to secondary for the double-tuned circuit, and thus assure the proper bandwidth for the required frequency response. The secondary winding of L9 is part of the 41.25-MHz sound trap. Coil L8 (41.25-MHz trap) develops a voltage at 41.25 MHz, and this voltage is added to the signal in the secondary of L9. When the signal voltage across the secondary winding of L9 is phased properly, the voltage across the trap winding (L8) acts to cancel the 41.25-MHz signal coupled from the primary to the secondary of L9. When trap L8 is properly aligned, the 41.25-MHz component of the signal present in the secondary of L9 is removed.

The first and second video i-f stages (the circuit shown in Fig. 4-9) are under agc control. An analysis of the circuit configuration of Q1 and Q2 shows the two transistors are connected in series with the B+ supply or source voltage. This circuit configuration effectively holds the current through both the transistors at the same level. The transistors used in this circuit have an inherent characteristic or property, in that their high frequency beta decreases as their collector current increases. This characteristic is utilized to establish agc control of these two stages by controlling or varying

the collector current. Due to the series-connected circuit configuration of the two stages, it is possible to shift the bias of one stage and indirectly shift the operating point of the other transistor. This changing bias action is accomplished by application of the agc voltage, which is varying in accordance with the strength of the received signal. In this manner the relative stage gain of the i-f amplifiers remains constant, independent of the strength of the received signal.

The overall i-f response of the receiver from the mixer to the video detector conforms with the response curve shown in Fig. 4-10. Note that this response fulfills the bandpass requirements for good color reproduction. These requirements are illustrated by the response curve shown in Fig. 4-8A.

The composite video signal appearing at the output of a video detector consists of the luminance, chrominance, sync, and color-burst signals. The photograph in Fig. 4-11 shows the waveform of the output signal for the period of one horizontal line. The drawing above the waveform depicts the color-bar picture on the screen at the time the

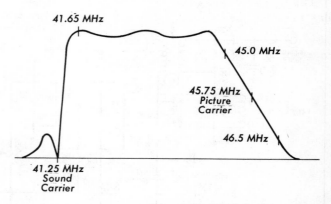

Fig. 4-10. Overall response of the rf and i-f sections of a color receiver.

waveform was taken. Note that the portions of the waveform that represent the colored bars consist of a 3.58-MHz chrominance signal superimposed on the luminance signal, whereas the white bar is represented by the luminance signal only. Although it is not evident from the waveform in Fig. 4-11, the portion of the chrominance signal that represents each bar has a specific phase relation to the color burst. This phase relation will determine the hues reproduced on the viewing screen.

The color-bar pattern shown in Fig. 4-11 was produced by an NTSC color-bar generator. This instrument has been replaced in modern color tele-

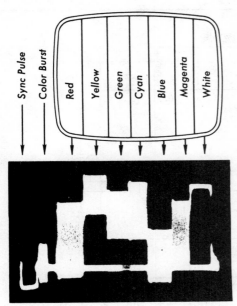

Fig. 4-11. Color-bar pattern and associated composite color at the output of the video detector.

vision servicing by the keyed-rainbow color generator discussed in Chapter 12.

The i-f stage of a color receiver must provide a flat response to the passband and also provide a fairly constant signal to the video detector. A change in the i-f response curve and extreme changes in signal level at the detector will affect the color seen on the screen of the receiver. An extremely good agc control system is a necessity in the color receiver.

Incorrect agc control can produce changes in the shape of the i-f response and also fail to compensate for signal-level changes. When this happens, the colors on the viewing screen will not be correct. For this reason, the agc circuit arrangements in most color receivers will have more involved circuitry than in the usual black-and-white receiver.

An earlier version of a video i-f section is illustrated in Fig. 4-12. This circuit employs vacuum tubes; however, circuit operation is very similar to that previously discussed. The first and second video i-f stages are stacked; that is, the plate current of the first stage is obtained from the cathode circuit of the second stage. This arrangement functions to assure balanced agc action for the first two video i-f stages.

The solid-state video i-f circuit discussed earlier used bipolar transistors. Many late-model color television receivers employ MOSFETs or inte-

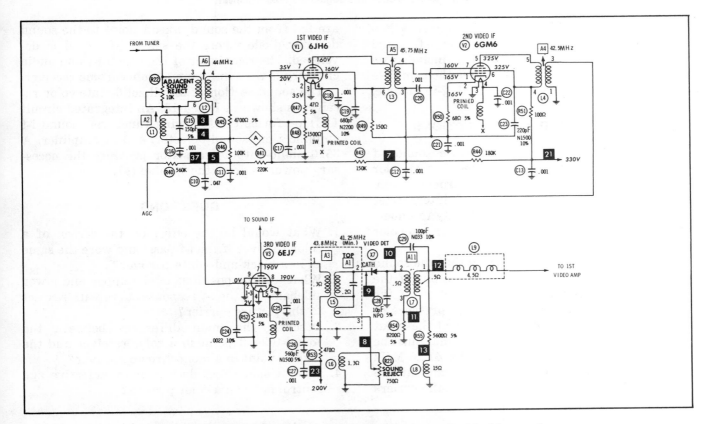

Fig. 4-12. Video i-f and detector circuits of a tube-type color television receiver.

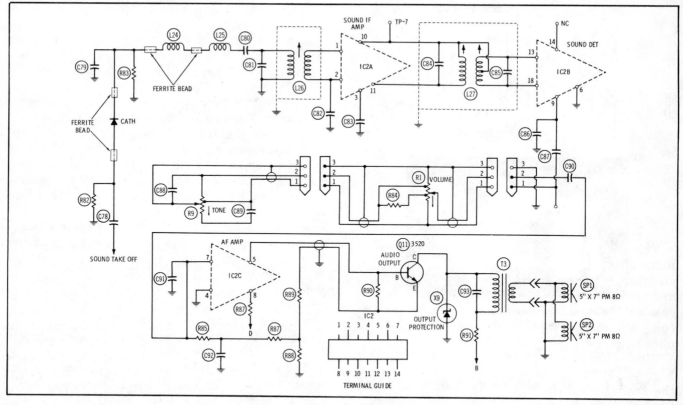

Fig. 4-13. Sound section of a solid-state color receiver.

grated circuits in their video i-f sections. When integrated circuits are used, the video i-f, sound i-f, agc, and sync circuits are often included on only two chips (ICs). In many cases the complete circuitry is mounted on a compact plug-in module.

SOUND I-F AND AUDIO SECTIONS

With the exception of the separate sound i-f detector, the sound i-f and audio sections of color receivers follow conventional monochrome design. If the reader knows the theory of intercarrier operation, he should have little difficulty in understanding and working with these sections in color receivers.

It has been previously mentioned that the sound carrier is severely attenuated at the input of the video detector circuit. Since the detection of the 4.5-MHz sound i-f signal requires the presence of both the video and the sound carriers at the input to the sound detector, the sound i-f signal must be obtained from a point ahead of the 41.25-MHz sound trap. This takeoff point is usually at the ouput of the final video i-f amplifier. Both carriers

are fed from the sound takeoff point to the sound detector diode where the sound i-f signal is developed. The remainder of the sound i-f and audio circuitry is the same as in a monochrome receiver.

The sound section of an all solid-state color receiver is shown in Fig. 4-13. An integrated circuit (IC) performs the following functions: sound i-f amplification, sound detector, and af amplifier. A 3520 audio-output transistor provides the necessary power to drive the speaker(s).

QUESTIONS

1. What would be the effect on the screen of a color receiver if the i-f passband were the same as in a black-and-white receiver?
2. What are the positions of the upper and lower sidebands on the i-f response curve with respect to the picture carrier?
3. Describe the main difference between the sound-takeoff point in a color receiver and the takeoff point in a monochrome receiver?
4. What visual effect does poor or defective agc control have on a color picture?

Chapter 5

Monochrome Circuitry

This chapter contains an analysis of the luminance channel, the agc circuit, the synchronization circuits, and the circuits that produce horizontal and vertical sweep voltages. Finally, there is a discussion about the operation of high-voltage circuits, and there is a description of the power supplies used in color receivers.

Block diagrams for these circuits are shown in Fig. 4-1. With the exception of the picture tube, the shaded blocks represent all circuits that are needed for monochrome operation. This conforms with the compatibility requirements for color television, because an important function of a color receiver is correct reproduction of monochrome pictures.

THE LUMINANCE CHANNEL

The main function of the luminance channel is to amplify the luminance portion of the video signal. This signal is comparable to a monochrome signal in that it represents the brightness variations of the image. From this standpoint, the function of the luminance channel can be compared with that of the video amplifier in a monochrome receiver. The luminance channel may be composed of one, two, or three stages to provide the desired luminance-signal level.

A secondary function of the luminance channel is to introduce a specific time delay in the luminance signal. This is necessary because all video signals undergo a time delay in reverse proportion to the bandpass limits of the circuits through which they pass. The time delay increases as the bandpass in narrowed. Since the luminance channel must pass a wider range of frequencies than the chrominance channel, the bandpass of the luminance channel is much wider than that of the

chrominance channel. Were it not for a special design feature of the luminance circuit, it would take a longer time for the chrominance signal to pass through the chrominance channel than it would for the luminance signal to pass through the luminance channel. The associated picture elements of these two signals must arrive at the picture tube at the same time; therefore, the luminance signal must undergo an extra time delay. This delay is accomplished through the use of a special delay circuit in the luminance channel. The characteristics of this delay circuit are covered in the following discussion.

A schematic of the luminance channel used in a tube-type color receiver appears in Fig. 5-1. The composite color signal at the input of the luminance channel is shown by waveform W1. This waveform shows the appearance of luminance, chrominance, and sync signals. Since all that is desired at the output of this channel is the luminance signal, takeoff points for the other signals are provided at the first video amplifier and the video cathode follower.

The takeoff point for the chrominance signal is in the plate circuit of the first video amplifier. Although not shown in Fig. 5-1, a 4.5-MHz trap located just after the video detector is used to prevent any remaining sound signal from appearing in the chrominance channel.

The video signal is fed to the grid of the video cathode follower through a peaking circuit consisting of L16 and a 6800-ohm resistor. The cathode-follower circuit is used to match the low input impedance of the delay line. Waveform W2 shows the video signal at the grid of the video cathode follower. This tube also functions as the sync amplifier. The sync signals are taken off from the plate circuit of V3A. The 3.58-MHz signal has been

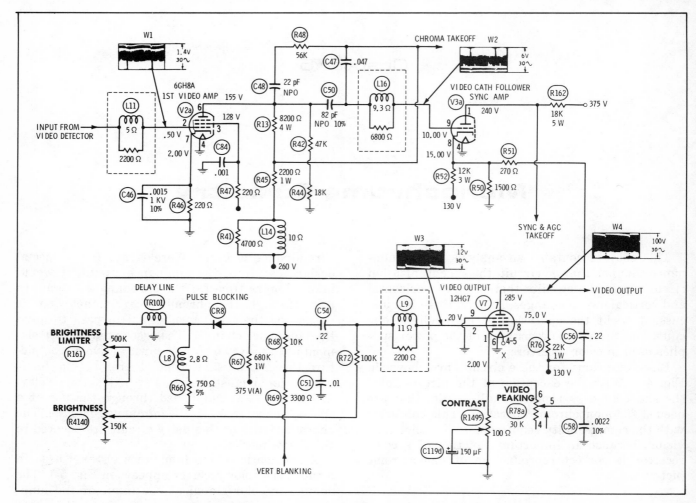

Fig. 5-1. Luminance channel and associated waveforms in a tube-type color receiver.

removed from the sync signal due to the reduced bandwidth of the luminance channel following the chroma takeoff point.

The luminance signal is taken off at the cathode of the video cathode follower. It is applied to the delay line through resistor R51. The physical appearance of the delay line used in this circuit is shown in Fig. 5-2A. Fig. 5-2B shows the delay line located on the topside of a solid-state chassis. The equivalent LC circuit of the delay line introduces a time delay of approximately one microsecond to the luminance signal. In order to minimize signal loss and to prevent standing waves, the characteristic impedance of the delay line must be matched at both ends. The value of the resistance at the input of the delay line is equal to the sum of R50 and R51, or 1770 ohms. The terminating resistance at the output of the delay line consists of L8 and R66, or 750 ohms.

The video signal at the output of the delay line is fed to the grid circuit of the video output stage

through diode CR8, capacitor C54, and a peaking network. The video signal at this point is shown by waveform W3. RC coupling is used to match the low impedance of the delay line to the high input impedance of the video output stage. Pulse-blocking diode CR8 is used to prevent the vertical blanking pulses from reaching the cathode circuit of the sync amplifier. The brightness control sets the grid bias for the video output stage and therefore determines the average brightness of the screen. The range of the brightness control is set by the brightness limiter control (R161).

The signal at the output of the luminance channel is shown by waveform W4. The video signal at this point is referred to as a negative Y signal; that is, maximum white is represented by negative-going peaks. The term Y signal is strictly a designation of luminance, or brightness, and therefore an increase in brightness is considered a negative change, or caused by a negative voltage excursion. Another way of remembering this po-

(A) Tube-type color receiver.

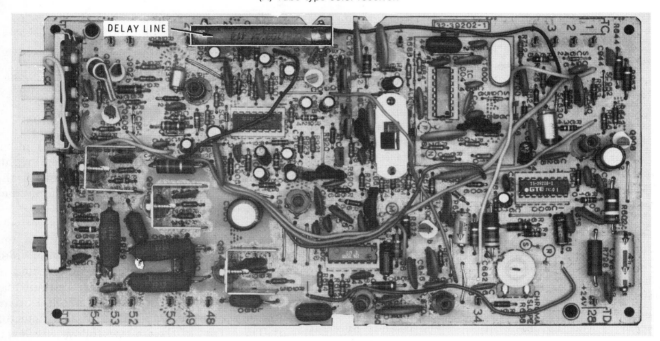

(B) Solid-state color receiver.

Fig. 5-2. Delay lines used in tube-type and solid-state color receivers.

larity designation is to keep in mind that a negative Y signal is of proper polarity to be fed to the cathode of the picture tube.

The output signal of the luminance channel in some receivers has a positive polarity. In this case, maximum white is represented by positive-going peaks.

With this exception, the operation of the circuit is essentially the same as that explained in the foregoing discussion. The luminance signal is amplified and properly delayed, unwanted signals are trapped out, and the necessary takeoff points are provided. Further discussion about the polarity

of the luminance signal will be presented when the matrix circuit is discussed.

Let us give further consideration to the delay line. The term *delay line* may be new to many technicians, yet all have worked with transmission lines and coaxial cables which are, in reality, forms of delay lines. In other words, the signal does not pass through these lines as fast as it does through space. Actually, a length of ordinary coaxial cable could be used as a delay line in a color receiver. Suppose that a length of RG-59/U were used. Since it is known that the amount of time delay needed for the signal is about 1 microsecond, the specific length of the cable can be calculated. The velocity of propagation through a length of RG-59/U is 66 percent of the speed of light. This means that a signal in this cable will travel at a speed that is 66 percent of the speed of a signal in space (or 186,000 miles per second). The time delay introduced by a cable can be stated as the time difference between the speed of a signal traveling across the terminal ends of the cable in space and the speed of the signal traveling through the cable.

Since the actual distance between the terminal ends of a delay line used in a color receiver is only a few inches, a straight conductor would introduce a negligible time delay. In order to calculate the number of feet of RG-59/U required to introduce a time delay of 1 microsecond, it is necessary to calculate the velocity of the signal through the cable. The speed through the cable is 66 percent of 186,000, or 122,760 miles per second. Thus, 648 feet of RG-59/U would be required to introduce a time delay of 1 microsecond.

Naturally, it would be impractical to supply 648 feet of cable with every color receiver, but a special cable can be designed so that the desired time delay will be obtained within a very short length. This is accomplished by increasing the inductance and capacitance of the conductor. Such a cable is used as a delay line in color receivers.

A drawing of the cable used as the delay line in an early color receiver is shown in Fig. 5-3A. Note that the signal-carrying conductor is wound around a flexible plastic tube. The fact that the conductor is coiled increases the inductance. The capacitance is increased when the conductor is coated with a mixture of powdered aluminum and styrene. These features will cause the velocity of propagation to be reduced so that a shorter length of cable can be used to introduce the necessary time delay. Eighteen inches of this cable is sufficient to produce a delay factor of approximately 1 microsecond. Delay lines such as the one shown in Fig. 5-3B have undergone even more reduction in size. The one shown here is slightly more than 5 inches long. This type of delay line is used in modern color receivers.

SOLID-STATE LUMINANCE CHANNEL

Most manufacturers are now using solid-state devices in color receiver circuitry. The requirements of the luminance channel circuit are such that it lends itself readily to solid-state devices, allowing the design engineer much more latitude.

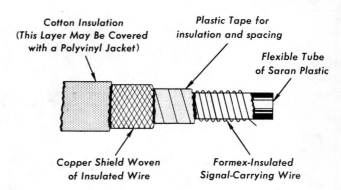

(A) Used in early color receivers.

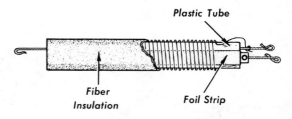

(B) Used in modern color receivers.

Fig. 5-3. Make-up of delay lines used in color receivers.

The circuit illustrated in Fig. 5-4 is used in an RCA CTC40 color receiver, and provides brightness limiting and video peaking along with the main function of amplification of the luminance component of the video signal.

The output of the video detector is effectively in series with the base bias of the first video amplifier. Base bias is developed by a voltage-divider network. The first video amplifier is connected in an emitter-follower configuration. This circuit features a high input impedance to match the inherently high output impedance of the detector circuit. The output of the emitter-follower is developed across a 1000-ohm resistor (R54) in the emitter circuit. Additional circuit loading results from coupling to the following stages. The output from the first video amplifier is connected to the first chroma amplifier circuit, the second video amplifier, the sync separator, the noise inverter, and the agc gate.

The signal output of the first video amplifier is coupled to the delay line. The delay line must be properly terminated to prevent ringing, faulty color registration, etc. The delay line used in this circuit has a characteristic terminal impedance (input and output) of 680 ohms at video frequencies. A 560-ohm resistor (R56) in combination with the first video amplifier output impedance provides the delay line with the proper 680-ohm input terminal impedance.

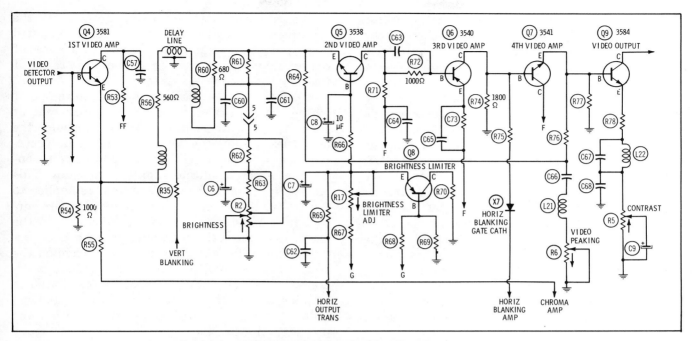

Fig. 5-4. Luminance channel of the RCA CTC40 color chassis.

The output terminal of the delay line is applied to the second video amplifier through a 680-ohm resistor (R60). The second video amplifier stage is designed to exhibit an ac input impedance of zero ohms. Therefore, the output of the delay line is effectively coupled to ac ground through a resistor to properly satisfy termination impedance requirements.

The second video amplifier utilizes a common-base configuration and is designed for an input impedance of zero ohms. This is accomplished through the use of a 10-μF bypass capacitor (C8). The stage functions as a power amplifier. Any fluctuations in the dc output of the first video amplifier are amplified throughout the range of video frequencies. This stage provides proper impedance matching between the delay line output and the third video-amplifier input. Positive-going pulses at the vertical rate are fed to the emitter of the second video amplifier to provide vertical retrace blanking.

The operating point of the second video amplifier varies with the setting of the brightness control. Any change in the brightness control results in a change of the operating point by changing the forward bias current. The lower the resistance of the brightness control, the greater the forward bias current. The result is a larger average current through the second video amplifier load resistance. This current is translated by the remaining video amplifiers as a reduction in crt cathode bias and,

consequently, an increase in brightness. The brightness control is ac-bypassed by capacitor C6.

The third video amplifier employs a pnp transistor in a common-emitter configuration. The video signal is fed to the base through a 100-ohm resistor (R72). This resistance provides proper loading for the second video amplifier, and impedance matching between the second and third video amplifiers. It also functions to prevent saturation of the third video amplifier in the event the second video amplifier develops a collector-emitter or emitter-ground short circuit. The output signal is developed across an 1800-ohm load resistor (R74) and is direct-coupled to the base of the fourth video amplifier.

The fourth video amplifier is connected as an emitter follower. The output of this stage is developed across a loading resistor and is direct-coupled to the base of the video output transistor. Positive bias voltage applied to the collector of Q7 is decoupled from the supply source by a filter network comprising a resistor and capacitor. This decoupling network prevents feedback loops that could cause low-frequency smear, etc. Horizontal pulses, which occur simultaneously with the horizontal retrace interval, are fed to the base to accomplish horizontal retrace blanking. This circuit operation is as follows. Horizontal pulses originating at the high-voltage transformer are applied through a clamp transistor to an isolation diode (X7). The isolation diode is reverse-biased during scan time by a positive dc voltage developed at the collector of the clamp transistor. During this interval, the blanking circuit is isolated from the fourth video amplifier to prevent the loss of high-frequency components of the video signal.

The negative-going horizontal pulses, fed to the isolation diode during retrace intervals, overcome the diode reverse bias and permit it to conduct. These negative-going pulses are present at the base of the fourth video amplifier and are of sufficient amplitude to effect cutoff. These pulses are applied to the crt through the video output stage. This action causes picture-tube cutoff (a dark screen) during horizontal retrace time.

A brightness-limiter circuit is employed in the CTC40 chassis to hold the crt beam current within proper limits. The drive potential of the horizontal deflection system used in this chassis is such that, with a high, nonlimited brightness control adjustment, it is possible to exceed the current capabilities of the crt.

The brightness-limiting action of the circuit functions to reduce the forward base-bias voltage on the second video amplifier when the preset limit of the crt beam current is attained. This preset limit is 1600 microamperes (1.6 mA). Circuit action is as follows. The high-voltage transformer secondary winding is returned to B+ through the brightness-limiter control. Therefore, all beam current drawn by the crt must pass through the brightness-limiter control. Connected between the low side of the brightness-limiter control and ground is the brightness-limiter transistor. The fixed base bias for this stage makes the voltage across it comparatively independent of the current through it, as long as it is conducting. This action is much like that of a zener diode, the zener voltage being determined by the resistive divider network in the limiter base circuit.

The current through the brightness-limiter control has two parallel paths—one through the brightness-limiter transistor, and the other through the crt. If the brightness control is adjusted in such a manner that the crt is cut off, the only path for current flow is through the brightness-limiter control and the brightness-limiter transistor. When the crt is cut off, this current will be 1.6 mA, the desired beam current limit. Should the brightness control be adjusted so that the crt starts drawing current, part of the current will flow through the crt and the remainder through the brightness-limiter transistor. The total current flow through the brightness-limiter control remains at 1.6 mA.

The constant voltage at the emitter of the brightness-limiter transistor supplies a regulated bias voltage of approximately 4 volts to the base of the second video amplifier throughout the range of the brightness-limiting system.

When the brightness control is set to the point where the crt draws the total preset current of 1.6 mA, all of the current flowing through the brightness-limiter control is beam current. Therefore, there is no current available to sustain conduction of the brightness-limiter transistor. This results in a loss of the constant voltage applied to the base of the second video amplifier. If more current is demanded by the crt, the voltage on the emitter of the brightness-limiter transistor decreases, reducing the forward bias voltage on the second video amplifier. This action results in a decrease of average conduction in the second video amplifier, and a decrease in brightness and crt beam current, holding beam current within the preset 1.6-mA limit.

The video-output circuitry is reminiscent of that previously employed in vacuum-tube circuit configurations. It consists of a common-emitter amplifier whose input is dc-coupled to the emitter of the fourth video amplifier, and whose output is dc-coupled to the crt.

The contrast control is used to vary the ac bypass of the series emitter resistance. The contrast control is ac-bypassed by a 30-μF capacitor. This circuit action effectively controls ac degeneration, with the end result being effective gain (or contrast) control.

Further control of the stage is provided by capacitor C68, which functions to reduce high-frequency degeneration and prevent changes in high-frequency response (peaking) at different contrast-control settings. Inductance and capacitance components in the emitter circuit form a 3.58-MHz trap which functions to reduce the effects of an objectionable interference pattern resulting from the mixing of chroma signals and high-frequency video signals.

THE AGC CIRCUIT

Although conventional in operation, the agc circuit plays an important part in the color receiver. This importance can be realized when it is considered that variations in the amplitudes of the incoming signal will affect the color as well as the brightness of the image. In order to stabilize the operation of the receiver, a good agc circuit is a necessity.

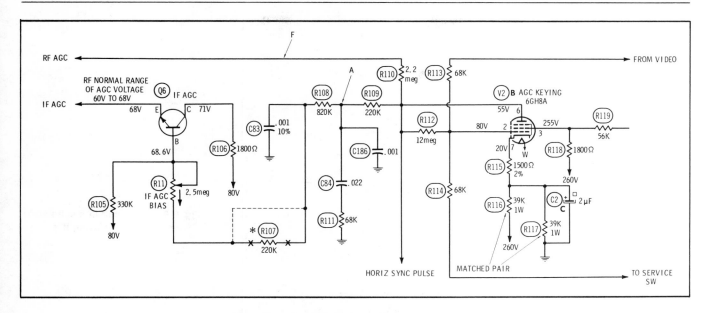

Fig. 5-5. Agc circuit employed in hybrid color chassis.

The schematic shown in Fig. 5-5 illustrates the agc circuit used in a hybrid color receiver chassis. The pentode section of a 6GH8A tube is connected as a keyed amplifier. The agc output voltage is driven more negative under conditions of increased signal strength. The voltage developed by the agc keyer circuit is applied to a transistor agc amplifier stage. The agc amplifier functions to translate the agc voltage to a corresponding current change. The output of the agc amplifier is then used to effect a shift in bias in the i-f agc amplifier circuit. The i-f agc amplifier is connected as an emitter follower whose output is developed across the emitter resistor of the first video i-f amplifier. When the received signal strength increases, the plate of the agc keyer tube will go more negative and the base of the i-f agc amplifier (normally biased form the 80-volt source through resistor R105) goes less positive. This results in decreased i-f agc voltage and the current through the first and second video i-f transistors increases sufficiently to maintain a constant voltage drop across their common source resistor (see Fig. 4-9). Therefore, i-f gain is decreased and the video-detector output remains relatively constant. The agc system functions to hold the video-detector output within limits over a wide variation in received signal-strength levels.

Many of the color television receivers now being manufactured employ a combination of discrete transistors and integrated circuits (ICs). The agc circuit used in one of these modern solid-state receivers is shown in Fig. 5-6. The agc gate, i-f agc amplifier, and rf agc delay stage are all contained in a single IC which includes the first and second video i-f stages. Keying pulses from the horizontal output transformer are fed to the agc gate through IC terminal 5, and video signals are fed to terminal 6 from the emitter circuit of the first video amplifier. The source voltage is provided through IC terminal 11, and the agc storage capacitor (C213) is connected to terminal 9.

The i-f agc amplifier provides amplification for the agc voltage, which is then fed to the video i-f stages and the rf agc delay circuit. A variable dc reference for the i-f amplifier is provided by i-f agc control R205. The collector voltage for rf inverter transistor Q207 is furnished by a voltage-divider network consisting of R216, R217, and R218. One end of the voltage divider is connected to the −21-volt source, and the other end is connected to the +23-volt source. The rf agc terminal is the junction of R217 and R218. The emitter of the npn agc inverter transistor is connected to ground through a 560-ohm resistor (R215). The conduction of the agc inverter transistor is determined by the voltage fed to its base by the rf agc delay stage through IC terminal 12. The rf agc control, R206, provides a variable dc reference for the rf agc delay stage. The agc inverter serves to invert and amplify the agc voltage applied to its base. Thus, a small voltage change at the base of Q207 will cause a larger change in the opposite direction at its collector. This voltage change, which also occurs at the junction of R217 and R218, is fed to the tuner as the rf agc voltage.

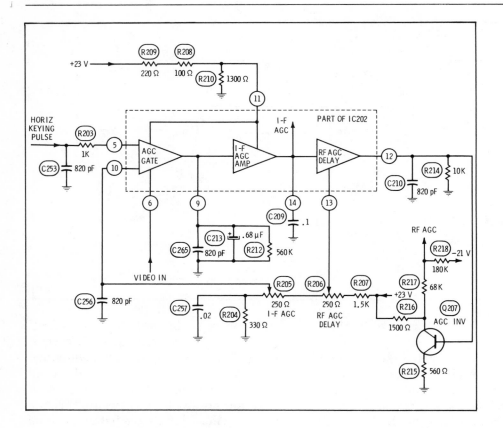

Fig. 5-6. Agc circuit in a solid-state color receiver using ICs.

SYNC CIRCUITS

The sync circuit in a color receiver performs the same function as it does in a monochrome receiver; it separates the synchronizing pulses from the video information. A simplified version of a sync separator is shown in Fig. 5-7.

The composite video signal applied to a sync separator is usually obtained from one of the video amplifiers or from a sync amplifier. The signal, composed of video information and synchronizing pulses, is applied to the grid circuit of V1 in Fig. 5-7. Positive-going sync pulses cause the sync-separator grid to draw current, producing a voltage drop across resistor R2. This bias changes according to signal level and permits only the sync pulses to be amplified. The amplitude of the video content is automatically maintained below the bias

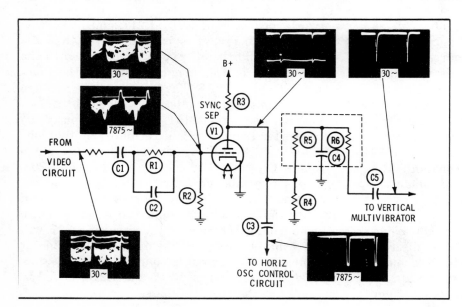

Fig. 5-7. Simplified version of a tube-type sync separator.

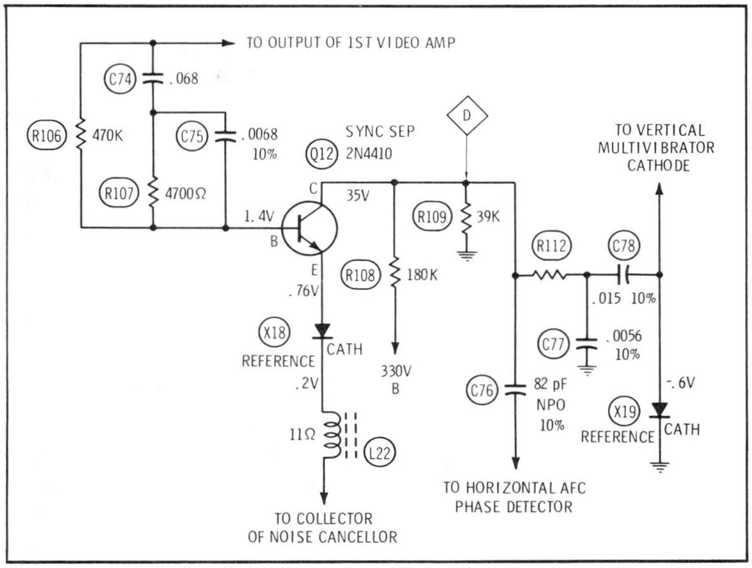

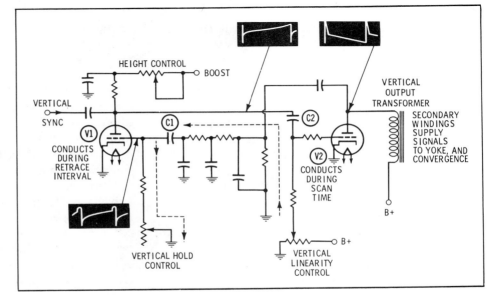

Fig. 5-8. Transistor-type sync-separator used in Sylvania D13 color chassis.

level and therefore cannot affect the plate current in tube V1.

The output of V1 contains only horizontal- and serrated vertical-sync pulses. The synchronizing pulses are applied to the horizontal-oscillator control circuit through capacitor C3.

The same train of pulses is also applied to the vertical multivibrator through an integrating circuit. The serrated vertical-sync pulses are shaped into a single pulse that occurs at the field scan rate. This vertical pulse is then used to synchronize the vertical multivibrator at the vertical sweep rate.

The schematic in Fig. 5-8 illustrates a transistor sync separator stage used in a Sylvania chassis D13. The basic circuit operation is the same as the description for the circuit illustrated in Fig. 5-7;

the transistor is biased so that only the sync pulses appear at its collector. However, a slight difference in waveform content can be expected in solid-state circuits.

VERTICAL SWEEP CIRCUITS

A modified form of a multivibrator is used to develop the vertical-sweep drive for a large number of tube-type television receivers. A simplified schematic of one such multivibrator is shown in Fig. 5-9. Two triodes are used for this circuit, and in most receivers, both triodes are usually contained in one envelope.

Triode V1 (Fig. 5-9) conducts during the retrace interval, and at this time capacitor C1 is

Fig. 5-9. Vertical-sweep multivibrator circuit used in a tube-type color television receiver.

charged by grid current from triode V1. This charge then keeps triode V1 cut off during the time required for the charge to leak off capacitor C1. This discharge path is shown by the dashed line. The length of time required for discharging capacitor C1 is controlled by changing the resistance of the vertical hold control.

When triode V1 is cut off, the increase in plate voltage of V1, coupled through capacitor C2 to the grid of V2, causes triode V2 to conduct. Current drawn through the primary of the vertical output transformer induces vertical-scanning and convergence voltages in the secondary windings.

When capacitor C1 is completely discharged, triode V1 begins conduction, and the drop in V1 plate voltage turns off triode V2. When the field in the vertical output transformer collapses, a high-amplitude positive pulse is applied to capacitor C1, causing it to charge by drawing grid current from triode V1. This starts a new vertical sweep cycle.

The vertical-sync pulses from the sync separator are applied to the plate of triode V1 and to the grid of triode V2. The sync pulses are negative going and applied during the time that triode V1 is cut off. Therefore, these pulses have no effect on triode V1, whereas triode V2 is conducting, and the negative sync pulse tends to reduce conduction in this tube. This action produces an amplified positive pulse at plate of V2, which is, in turn, applied through capacitor C1 to the grid of V1. Triode V1 is turned on, and this triggers a new vertical-sweep cycle.

The schematic of a solid-state vertical-sweep section is illustrated in Fig. 5-10. For purposes of explanation this is a simplified illustration.

The fundamental vertical-sweep system employed in the RCA CTC40 chassis is illustrated in Fig. 5-10A. The integrator sweep circuit consists of a high-gain amplification system operating in conjunction with an integrating capacitor. Operation is as follows.

At the start of vertical trace, the integrating capacitor, C1, is charged from a voltage source. This capacitor charge causes the amplifier to supply yoke current, resulting in a voltage being developed across the feedback resistor, R1, which is coupled directly to the integrating capacitor. This feedback action maintains the amplifier input voltage at a constant level, producing a constant rate of voltage "build up" across the integrating capacitor. The voltage developed across the feedback resistor is directly proportional to the yoke current. Therefore, the increase of the yoke current is constant, and a linear scan is produced.

The vertical-sweep rate is determined by an electronic switch which discharges the integrating capacitor at a 60-Hz rate. Vertical-sync pulses are applied to the switching transistor and determine the exact instant the switch is pulsed "on." This action synchronizes the vertical switching action with the transmitted vertical scanning interval. The "linearity clamping" transistor provides the initial charging current to the integrating capacitor.

The function of the vertical switch (Fig. 5-10B) is to provide a discharge path for the integrating capacitor at the end of each vertical scan interval. This action causes beam retrace and prepares the circuit for the next vertical scan function. Operation of the vertical switch is made self-sustaining by the action of two feedback paths. One path, consisting of resistors R1 and R2 and capacitor C1, is applied to the base and provides the appropriate pulse to initiate "turn-on." Vertical-sync pulses from the sync separator are integrated by resistors R3 and R4 and capacitor C2, and add to the triggering waveshape. An additional feedback voltage is applied to the switch from the vertical output transformer via the vertical hold control. This additional voltage causes the switch base to pass rapidly through the "turn-on" voltage potential. As a result, switch "turn-on" is extremely stable and comparatively immune from random noise pulses. The vertical hold control has some control of the "turn-on" point and, therefore, the frequency at which the circuit operates.

Since it is necessary to provide a sufficient amount of initial charging current for the integrating capacitor, a special clamping circuit called the *linearity clamp* (Fig. 5-10C) is utilized. Operation of this circuit is as follows.

The action of the vertical-switch discharging capacitor, C1, also cuts off the predriver transistor. This produces a positive voltage pulse on the collector of the predriver. This voltage is of sufficient amplitude to forward-bias the linearity clamp transistor. The linearity clamper conducts and causes a current through the transistor via R1 and the vertical switch. The vertical switch turns off after approximately 700 microseconds, and the linearity clamp current then rapidly charges capacitor C1. As the charge rapidly builds up on capacitor C1, the predriver and driver stages start to conduct, causing the linearity clamp base-emitter junction to become reverse-biased due to the less-positive voltage on the collector of the predriver. This circuit action cuts off the linearity clamp and initiates vertical scan. Capacitor C1

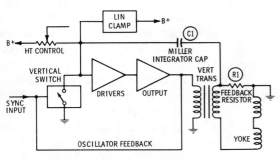

(A) Integrator sweep circuit.

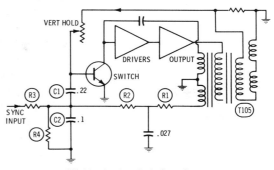

(B) Vertical-switch function.

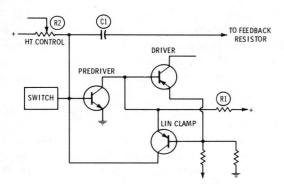

(C) Vertical-linearity clamp.

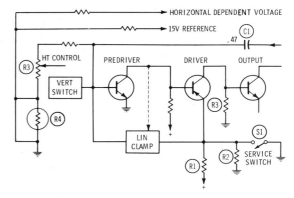

(D) Vertical predriver, driver, and output circuits.

(E) Vertical-output circuitry.

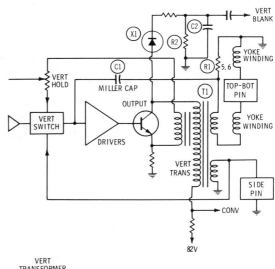

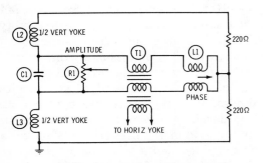

(F) Top-bottom pincushion correction.

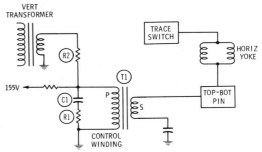

(G) Side pincushion correction.

Fig. 5-10. Simplified illustration of a solid-state vertical-sweep section.

continues to charge through the height control, R2, for the duration of scan time.

The vertical driver section (Fig. 5-10D) is much more familiar. It consists of two stages—a predriver (npn transistor operating as a common-emitter amplifier), directly coupled to a driver (pnp transistor operating as a common-emitter amplifier). Emitter supply voltage for the driver stage is obtained from a voltage-divider network composed of R1 and R2. The driver collector load comprises R3 and the base-emitter junction resistance of the vertical output stage.

Provisions for picture-tube screen setup are provided by switch S1, which functions to "short" the driver emitter to ground when actuated. The waveshape of the input signal to the predriver is determined by the charging action of the integrator capacitor C1, which is charged through the height control, R3. The height-control supply voltage is made relatively immune to temperature-induced variables by the action of thermistor R4. A degree of dynamic regulation for the circuit is provided by a signal from the horizontal-deflection system. The insertion of this voltage tends to maintain a constant vertical sweep or height, regardless of horizontal scan and high-voltage fluctuations.

The function of any vertical-output circuit is to provide the power necessary to fulfill the vertical deflection requirement of the crt beam. In the RCA CTC40 chassis, the vertical-output stage is a common-emitter amplifier with an input from the driver stages (Fig. 5-10E). Loading for the vertical-output stage is provided by the vertical-output transformer, T1, and the vertical-convergence circuit.

The vertical-output transformer is loaded by the vertical windings of the yoke, two feedback networks, and the pincushion-correction circuit. Integrating capacitor C1 is connected to the junction of resistor R1 and the secondary winding of the vertical-output transformer. There are two feedback networks connected to the vertical-switch transistor from the vertical-output circuit; both of these networks perform waveshaping functions to provide stable, self-sustaining vertical switching. Diode X1, in conjunction with capacitor C2 and resistor R2, provides a protective clamping action for the vertical-output transistor. Positive-going retrace voltage pulses cause diode X1 to conduct, effectively clamping the vertical-output collector to the voltage existing across capacitor C2. A relatively slow discharge path for capacitor C2 is provided by resistor R2. This discharge action sufficiently reduces the voltage across C2 during retrace time to ensure the necessary voltage difference across diode X1 when retrace pulses occur. The pulses that appear across capacitor C2 during conduction of X1 are applied to the second video stage to provide vertical-retrace blanking.

Top and bottom pincushion correction in this chassis is accomplished in a manner similar to methods used in tube-type color chassis. A signal voltage derived from the horizontal-yoke circuit is coupled to transformer T1 (Fig. 5-10F) to energize a circuit composed of capacitor C1 and coil L1, which is tuned to 15,750 Hz and is in series with the vertical-yoke windings, L2 and L3. The resultant sine wave is added to the vertical-yoke current waveshape in the proper phase and amplitude to effectively correct top and bottom pincushion distortion. A limited amount of control over the correcting sine-wave phase and amplitude is provided by variable inductor L1 and the damping resistance of R1.

Side pincushion correction (Fig. 5-10G) is accomplished by amplitude modulation (at a vertical rate) of the horizontal deflection current. This produces an increase in horizontal-scanning width at the center of the raster, with respect to the width at the top and bottom. This operation is made possible through the utilization of the saturable-reactor circuit illustrated in Fig. 5-10G.

A parabolic waveshape occurring at the vertical frequency is initiated by the action of the control winding of transformer T1, capacitor C1, and resistors R1 and R2. This waveform, coupled to the horizontal-yoke circuit by transformer T1, modulates the amplitude of the horizontal-yoke scanning current, producing the proper change in raster width.

HORIZONTAL OSCILLATOR AND AFC

The horizontal-control circuits for the color receiver do not differ from those used in the monochrome receiver. The schematic of Fig. 5-11 is the horizontal oscillator and afc used in a tube-type color receiver. The sync pulse from the sync separator and a sawtooth signal from a winding on the horizontal-output transformer are compared in a phase-detector circuit composed of two semiconductor diodes. A change in phase between these two pulses develops a correction voltage that is applied to the grid of the first triode. This voltage change is amplified and applied to the wiper contact of the horizontal hold control. The setting of this control establishes the dc bias at the grid of

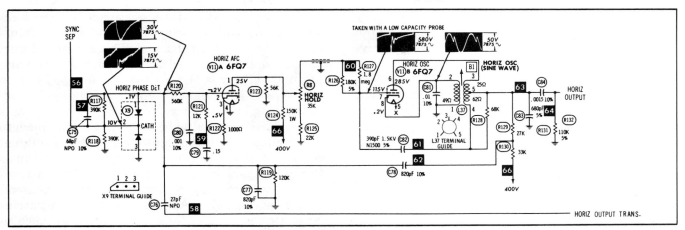

Fig. 5-11. Horizontal oscillator and afc circuit used in tube-type color receiver.

the oscillator and determines the free-running frequency of the oscillator. The correction signal varies the voltage at the wiper arm of the horizontal hold control and continually adjusts the oscillator frequency so that it matches the horizontal line-scanning frequency of the incoming signal.

The output of the horizontal oscillator is used to drive the horizontal-output stage. In a few receivers a driver, or buffer, stage is used to isolate the oscillator from the horizontal-output stage.

New concepts in circuit design and manufacturing techniques are currently being employed by manufacturers of color receivers. With the utiliza-

tion of solid-state devices, circuits are being designed and employed that exert even more exacting frequency control. The schematic in Fig. 5-12 illustrates the circuit used in an RCA solid-state color chassis. Circuit operation is as follows.

The function of an automatic frequency control (afc) circuit in the horizontal section of a television receiver is to hold the horizontal-oscillator frequency at exactly the same frequency and phase as the transmitted horizontal-sync pulses. Color television receivers demand even more exact performance from the horizontal afc circuit because the color-burst amplifier is keyed by pulses ob-

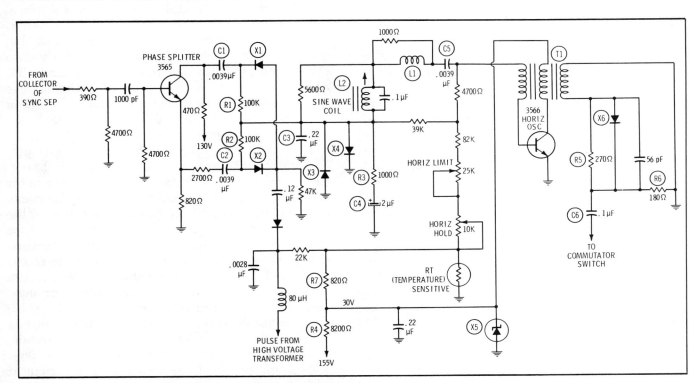

Fig. 5-12. Horizontal oscillator and afc circuit used in RCA solid-state color chassis.

tained from the horizontal-sweep circuit. Any change in frequency or phase existing between the occurrence of color burst and the horizontal-keying pulse for the burst amplifier will result in an incorrect color display.

The design of this afc circuit is such that it will hold the horizontal oscillator at the exact frequency of the incoming horizontal-sync pulses as long as the free-running frequency of the oscillator is within a tolerance of ±300 Hz of the transmitted horizontal-sync signal.

The schematic for the horizontal afc and oscillator circuitry is shown in Fig. 5-12. The phase-splitter stage supplies equal and opposite sync pulses to the dual-diode phase detector. Incoming sync pulses are differentiated at the base of the phase-splitter transistor to reduce possible interfering from the vertical-sync pulses that are present in the output of the sync separator.

Output pulses from the collector and emitter of the phase splitter are coupled to the phase-detector diodes by capacitors C1 and C2.

A reference pulse taken from the high-voltage transformer is applied to the junction of the diodes through a waveshaping network. This network shapes the negative-going pulses from the high-voltage transformer into a sawtooth signal suitable for application to the afc circuit. The frequency of the pulses sampled from the high-voltage transformer is the same as that of the horizontal oscillator.

When the pulses from the high-voltage transformer and the incoming horizontal-sync pulses occur at exactly the same frequency, each diode is keyed into conduction by the sync pulses as the reference voltage passes through zero. The current through each diode will be equal, resulting in equal and opposite charges on capacitors C1 and C2. As these capacitors discharge through resistors R1 and R2, equal and opposite voltages are developed across the resistors. The voltage at their junction is zero (with respect to ground) ; consequently, the amount of correction voltage developed is zero.

If the horizontal oscillator is running at a frequency less than that of the incoming horizontal-sync pulses, a change in the relative position of the reference voltage waveshape during the application of the sync pulses will result. Sync pulses will key the diodes into conduction during the positive portion of the retrace slope. Diode X1 will conduct more strongly than diode X2, and the charge on capacitor C1 will become more positive, while the charge on capacitor C2 will become less negative.

Discharge action of these capacitors through resistors R1 and R2 will result in an imbalance of current flow through R1 and R2, and the voltage developed at their junction will go positive. This positive voltage is the correction voltage for the horizontal oscillator and will cause the oscillator to increase in frequency.

Should the oscillator be running at a frequency greater than the incoming sync pulses, a negative correction voltage will be developed at the junction of R1 and R2 as the result of circuit action similar but opposite to that described in the preceding paragraphs. Application of the negative correction voltage to the oscillator will produce a decrease in the oscillator frequency.

The dc correction voltage present at the junction of R1 and R2 is fed to an afc limiting and filtering circuit comprising diodes X3 and X4, capacitors C3 and C4, and resistor R3. The function of this circuit is twofold—the limiting diodes prevent the afc correction voltage from exceeding −0.5 V to +0.5 V, and the filter network prevents the afc output from being contaminated by unwanted frequencies, such as 60 Hz. (A 60-Hz signal present at this point would result in horizontal pulling, twisting, etc.)

A blocking-oscillator circuit is employed as the horizontal oscillator in this circuit. Basic circuit action is as follows. Voltage pulses present on the collector of the horizontal-oscillator transistor are transformer-coupled into the base circuit, driving the stage into cutoff. During the time that the oscillator transistor is cut off, capacitor C5 discharges through the horizontal-hold control circuitry until the "turn-on" potential of the oscillator is reached. The oscillator conducts, a pulse appears at the collector and is coupled to the base, and the cycle repeats.

The adjustments of the horizontal-limit and horizontal-hold controls determine the discharge time of capacitor C5 or, in other words, the length of time the transistor remains cut off. In this manner, the horizontal hold control determines the frequency of the horizontal oscillator.

This is accomplished by adding the correction voltage from the afc circuit to the charge on capacitor C5. This correction voltage, depending on its polarity and amplitude, will either add to or subtract from the charge on capacitor C5 which, in turn, either increases or decreases the time required to discharge C5 to the turn-on potential of the oscillator. This circuit action alters the frequency and phase of the oscillator in accordance with the broadcast sync pulse.

HORIZONTAL-OUTPUT AND HIGH-VOLTAGE CIRCUITS

The basic design of the horizontal-output and high-voltage circuits used in a color receiver follows the design used for monochrome receivers; however, more exacting operation is required of these circuits in a color receiver. For one thing, the ultor voltage (the second-anode voltage) of the picture tube may have a potential of 30 kilovolts or greater. In addition, this voltage must be maintained under operating conditions which are continuously varying. Other dc voltages of considerable potential must also be furnished to some types of picture tubes by the high-voltage supply. These voltages are for the focus and the convergence anodes. The purposes of these voltages will be explained fully when the discussion of the picture tube is presented.

The schematic of the horizontal-output, high-voltage circuits of a tube-type color receiver is shown in Fig. 5-13. Sweep voltages to the yoke are supplied in a conventional manner by tapping the primary winding at terminals 3 and 4. The focus voltage is obtained by rectifying a pulse taken from a tap on the primary winding of the horizontal-output transformer. This voltage is developed in the same manner as the ultor voltage, except that a 66-megohm bleeder resistor (R116) is connected across the output of the 2AV2. The focus control is an adjustable coil that applies a positive pulse to the cathode of the focus rectifier through a 130-pF capacitor (C111). This positive voltage on the cathode of the focus rectifier reduces the effective potential between its plate and cathode, thus causing the focus voltage to be reduced. The amplitude of the positive pulse is varied by adjusting the slug in the coil. This arrangement permits the focus voltage to be adjusted for the best possible focus of the picture.

The ultor voltage is derived conventionally except that the output is regulated by a special high-voltage regulator tube and associated circuitry. Voltage regulation is accomplished through the use of the 6BK4 high-voltage triode which is connected between the high-voltage output and ground. This type of circuit is known as a shunt regulator. The bias on the grid of the 6BK4 is established by a voltage-divider network consisting of R118, R122, and R169. In many color receivers, a potentiometer is used in this circuit as a high-voltage control. The setting of the potentiometer determines the grid bias on the shunt regulator and, therefore, the current through the tube.

As the current through the shunt-regulator increases, the high voltage decreases due to the loading effect of the tube. In this manner, the high voltage can be set to a predetermined value.

During the time of an all-white picture, the beam currents in the picture tube are considerably high and very little current flows through the regulator tube. When an all-black picture is transmitted, there are no picture-tube beam currents and the shunt regulator conducts heavily. This change in regulator current is brought about as follows. When the beam currents increase, the high voltage starts to decrease because of the additional load. This causes a reduction in the instantaneous value of the boost voltage which is applied to the grid of the regulator tube through a .01-μF capacitor. As a result, the voltage on the grid of the regulator tube starts to decrease and the current through the tube decreases. When the crt beam currents decrease, the voltage on the regulator grid starts to increase. As a result, the current through the tube increases and the high voltage is held relatively constant.

A simplified circuit of a high-voltage regulator is shown in Fig. 5-14. Notice that the bias is shown as 20 volts. If the boost source is lowered, the 390 V at the grid of the regulator will be lowered. The difference between the grid and cathode voltages will be greater, and this bias will tend to reduce the current through the regulator. This action tends to keep the ultor voltage at the designated 25 kilovolts.

It has been stated that an increase in beam currents is accompanied by a decrease in current through the regulator tube. Also, a decrease in beam currents is accompanied by an increase in current through the regulator tube. Voltage regulation is accomplished because the current drain on the high-voltage supply does not change, and the voltage therefore remains constant.

An additional feature of the high-voltage regulator shown in Fig. 5-13 is the hold-down circuit, sometimes called a "fail-safe" circuit. Normally, the voltage on the control grid of the 6LQ6 horizontal-output tube is −55 volts due to grid rectification of the signal from the horizontal oscillator. An additional negative voltage is applied to the 6LQ6 grid through R165, CR107, R187, and CR106. This voltage is developed across R186 due to the rectification of a pulse applied to diode CR103 from the horizontal output tube. A positive voltage from the cathode of the 6BK4 regulator is applied to the 6LQ6 grid through an 8.2-megohm resistor (R173). This positive voltage nearly can-

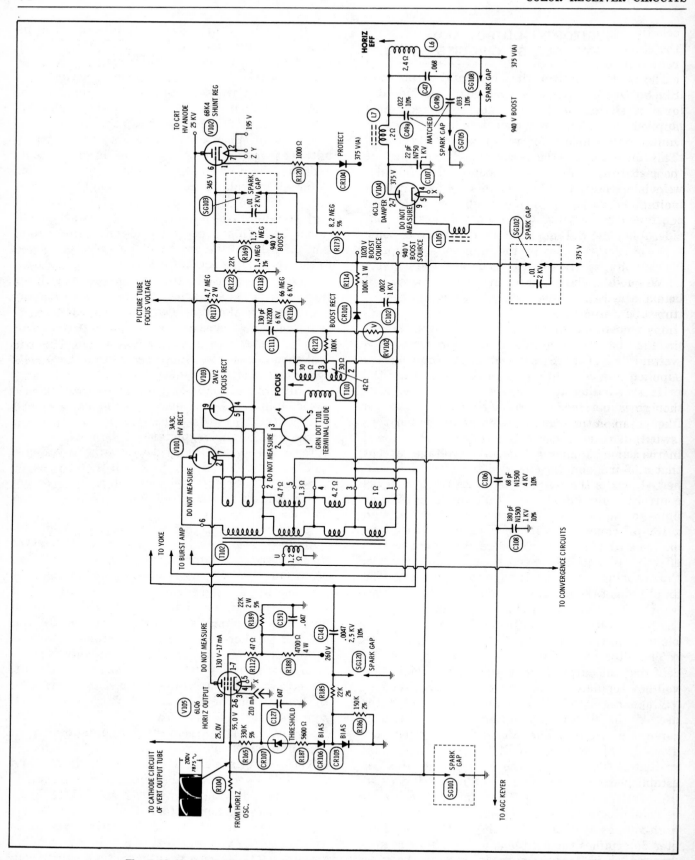

Fig. 5-13. Horizontal-output and high-voltage circuits used in a tube-type color receiver.

cels the negative voltage applied through R165. If the shunt regulator fails, the loss of cathode current will turn off diode CR104 in the cathode circuit and the cathode voltage will drop to zero. Since there is no longer a positive voltage at the grid of the 6LQ6 to cancel the negative voltage applied through R165, the conduction of the horizontal-output tube will be significantly decreased. This will cause a reduction of the high voltage and poor picture quality. Almost all late-model color television receivers employ some type of high-voltage hold-down circuit.

SOLID-STATE HORIZONTAL-OUTPUT AND HIGH-VOLTAGE CIRCUITS

When solid-state devices are used to attain the same end result as that achieved with vacuum tubes, different circuit arrangements and functions are often utilized. The schematic illustration in Fig. 5-15 is the horizontal-output and high-voltage circuit used in an all solid-state chassis. Operation of this circuit is as follows.

It is necessary to slightly alter the shape of the horizontal-oscillator output waveform to minimize the possibility of pretriggering the commutator switch. This is accomplished by the waveshaping network composed of diode X6, capacitor C6, and resistors R5 and R6 (see Fig. 5-12).

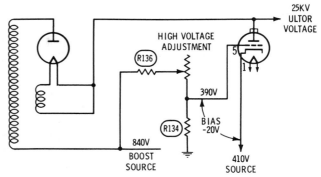

Fig. 5-14. Simplified high-voltage regulator circuit.

The voltage developed across the output winding of the horizontal blocking-oscillator transformer is coupled to this waveshaping network. Resistor R6 and capacitor C6 function as a differentiating network, producing a positive voltage spike to turn on the commutator switch. The diode, X6, is reverse-biased during the negative-going portion of the output voltage waveshape. This permits capacitor C6 to discharge through the parallel paths provided by R5 and R6. This discharging action holds the waveshape negative until the next positive pulse arrives. Thus, the commutator gate is held negative during trace time, reducing commutator pretriggering.

It is an inherent characteristic of transistors that their operation will vary with changes in

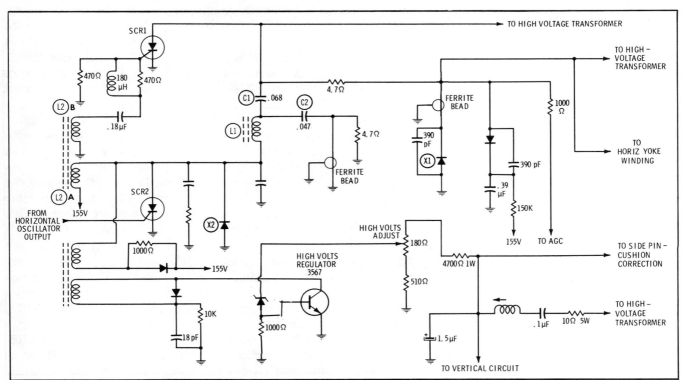

Fig. 5-15. Simplified schematic of solid-state horizontal-output and high-voltage circuit.

ambient and internal temperature. A thermistor, RT, in conjunction with resistor R7, functions as a temperature-sensitive, voltage-divider network. As the temperature of the horizontal-oscillator transistor changes, its operating frequency tends to change.

The same changes in temperature that affect the transistor also affect the thermistor (RT). The transistor base-circuit voltage will be altered by the temperature-induced changes in the divider network consisting of RT and R7. This change in base-circuit voltage will be in a direction that will cancel out the effects of temperature on the transistor.

The RCA CTC40 chassis utilizes two silicon controlled rectifiers (SCRs) and their associated components to generate the necessary yoke current and fulfill high-voltage requirements.

The function of any horizontal-deflection system used in television receivers utilizing electromagnetic deflection is to provide a linear current through the yoke windings. In turn, this current moves an electron beam from one side of the picture-tube screen to the other in a linear sweep. This action is normally referred to as *trace,* and the yoke current that caused the deflection is called *trace current.*

Trace current must be in sync with the incoming television signal. The yoke current must also provide a means of returning the crt beam to the starting side of the crt screen. The current that accomplishes this is referred to as retrace, or flyback, current.

A partial schematic of the horizontal-output circuit is shown in Fig. 5-16. Diode X1 and silicon controlled rectifier SCR1 control the current through the horizontal-yoke windings during the crt beam trace time. Diode X2 and silicon controlled rectifier SCR2 control the flow of current through the horizontal-yoke windings during retrace time.

Energy storage and timing properties are provided in the circuit by components L1, C1, C2, and Cy. Inductors L2A and L2B provide a charge path for L1 and C1, and a gating, or keying, signal to SCR1. The complete horizontal-deflection yoke-current cycle can be divided into a sequence of individual actions involving different modes of horizontal circuit operation. These actions are accomplished during discrete intervals of the horizontal-deflection yoke-current cycle.

During the first half of crt beam trace time, the current through the horizontal-deflection coils decreases toward zero and flows through trace diode

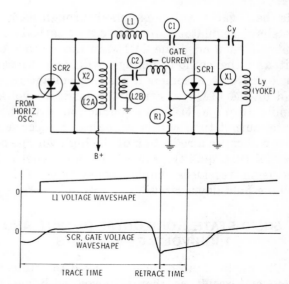

Fig. 5-16. Simplified schematic of the horizontal-output circuit shown in Fig. 5-15.

X1, resulting in a charge buildup on capacitor Cy. During this interval (first half of trace time), silicon controlled rectifier SCR1 (the trace SCR) is prepared for conduction by the application of the proper gate-voltage pulse. However, SCR1 will not conduct until its anode-cathode junction is forward-biased. This condition will be satisfied during the second half of the beam trace cycle.

At the end of the first half of the trace, the yoke current reaches zero, capacitor Cy starts discharging through the yoke inductance, and the current through the circuit reverses, reverse-biasing diode X1 and, simultaneously, forward-biasing SCR1. The capacitor discharges into the yoke inductance through SCR1, and the resulting yoke current completes the second half of trace.

When the second half of the trace is concluded, the crt beam has scanned across the entire width of the crt screen. At this point, a pulse, derived from the horizontal-oscillator circuit, keys the retrace SCR into conduction. This action releases the charge previously built up, or stored, on capacitor C1, and current flows into the commutator circuit comprising inductor L1 and capacitor C1.

Because of heavy forward current flow through the yoke circuit (SCR1, Ly and Cy), the net current resulting from the combined circuit actions of the commutating switch circuit and the yoke circuit continues to allow the trace rectifier, SCR1, to conduct.

At this point both rectifiers, SCR1 and SCR2, are conducting. However, the current in the commutator circuit increases much more rapidly than the current in the yoke circuit. After an extremely short period of time (two or three microseconds), the net current in SCR1 reverses, turning off SCR1 at the start of retrace.

Circuit conditions are now set to initiate retrace (see Fig. 5-16). Trace rectifier SCR1, along with diode X1, is cut off, and retrace rectifier SCR2 is conducting. The result is a series-resonant circuit consisting of inductor L1, capacitor C1, and the horizontal-yoke windings. (Capacitor Cy is also in series with these components but, because of its value, can be disregarded.)

The current through this circuit causes the crt beam to retrace halfway across the screen. At this point the current flow has decreased to zero. Current flow in the series-resonant circuit now reverses, and retrace rectifier SCR2 ceases conduction because the current flow in the circuit is opposite the normal flow of forward current.

Diode X2 is now forward-biased by this reversal of current and starts conducting, supplying the energy for the remainder of retrace. The energy previously stored on capacitor C1 has been returned to the yoke inductance.

Retrace current in the horizontal-yoke winding returns the electron beam to its starting point. The time interval of yoke retrace current is made equal to the desired retrace time by selection of the proper values of components L1, C1, and Ly.

These components are selected to be resonant at a frequency that has a period equal to two times the retrace time interval. Therefore, the current during one-half cycle of circuit oscillations will accomplish the full retrace function.

After completion of one full cycle (trace and retrace), the circuit must be made ready for the next cycle. This includes restoring energy to the commutator circuit and resetting the trace rectifier, SCR1. Both of these functions are performed by utilizing circuitry which includes inductor L2.

During retrace, inductor L2A is connected between B+ and ground by the conduction of SCR2 and diode X2, respectively. When X2 ceases conduction, inductor L2A is removed from ground. A charge is built up on C1 from the B+ line through inductor L2A. This charging process continues throughout the trace interval, until retrace begins. The charge on C1 serves to replenish energy to the yoke circuit during the retrace interval.

The voltage developed across inductor L2A during the charging of capacitor C1 is used to forward-bias the gate of SCR1. This sets up SCR1 and enables it to conduct upon receiving the proper signal. The voltage developed across inductor L2A is coupled to the gate of SCR1 via L2B, C2, and R1.

These components form a waveshaping network that forms a pulse with the proper shape and amplitude to enable SCR1 to conduct when its anode-cathode junction is forward-biased. This will occur approximately midway through the trace interval.

SONY SOLID-STATE HORIZONTAL-OUTPUT STAGE

The horizontal-deflection circuitry used in many Sony color receivers differs from the conventional circuits used in other receivers. The functions normally performed solely by the horizontal-output stage are divided into two sections. Division of the functions has several advantages. Two moderately high-power transistors can be operated well within their maximum ratings to provide reliable performance at a lower price than the special high-power transistor normally required. By separating the high-voltage generation and horizontal-deflection functions, the deflection current is made independent of the crt beam current. Also, the pulse width of the high-voltage supply is no longer restricted to the retrace time. Because of this the high voltage can be generated with a long-duration, low-amplitude pulse, which imposes less stringent requirements on the power transistor with regard to collector breakdown voltage.

The horizontal-deflection system (Fig. 5-17) is the simpler of the two branches. This section contains only the horizontal-output stage. The horizontal-deflection coils (HDY) and output transistor Q801 are connected to the same point across the horizontal-output transformer (HOT). To minimize their total inductance, the deflection yoke coils are connected in parallel. This in turn minimizes ringing and reduces the peak amplitude of the flyback pulse and the possibility of exceeding the collector breakdown voltage of transistor Q801. Since the horizontal-size coil (L601) is in series with the deflection coils, varying the coil inductance changes the deflection current, and hence the sweep width. Diode X801 is the damper diode. The flyback pulse is rectified by diode X803. The rectified voltage is effectively in series with the 90 volts supplied to Q801. This total voltage is used as the focus and screen-voltage supply for the crt. Secondary windings on the horizontal-output transformer supply sweep voltage to the tilt control, the afc, and the blanking circuits. This

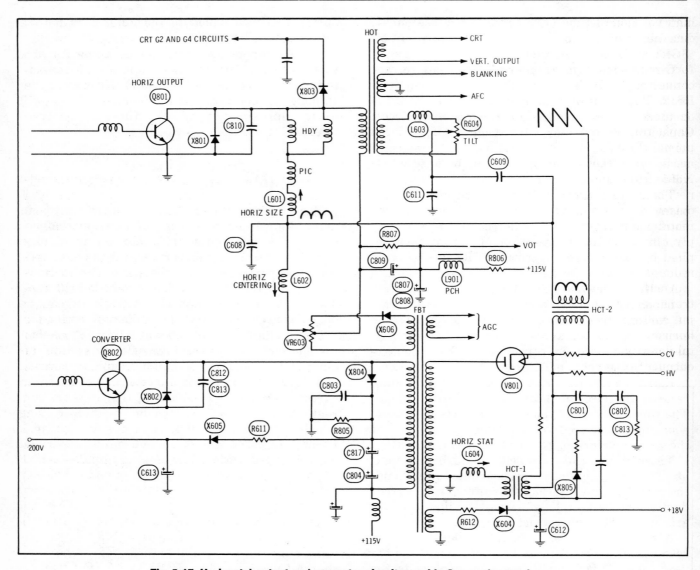

Fig. 5-17. Horizontal-output and converter circuits used in Sony color receivers.

sweep voltage applied to the tilt control (R604) is connected to a sawtooth waveform by L603. The parabolic voltage across C608 and the sawtooth waveforms are combined at HCT-2.

The horizontal-converter section (Fig. 5-16) is quite elaborate. This section furnishes the dc operating voltages for the crt and related circuits, and also supplies the low voltage for most of the other circuits of the receiver.

The ac power to supply the various rectifiers is generated by transistor Q802, diode V802, and flyback transformer FBT. A clipper to protect Q802 from transient pulses generated by high-voltage arcing is formed by diode X804, resistor R805, and capacitors C803, C804, and C817. When a high-voltage arc occurs, the voltage transient that is generated is coupled back to the primary of the

flyback transformer. Its amplitude may be greater than the collector breakdown voltage of Q802. However, the capacitors in the clipper circuit are charged to the peak value of the normal flyback pulse. A higher-amplitude pulse appearing at the collector of Q802 will forward-bias X804, and then be absorbed by the capacitors.

High-voltage rectifier tube V801 is supplied with a high-voltage ac by a winding on the flyback transformer. A tickler voltage on the flyback transformer provides the filament voltage for V801. The full output of V801 supplies 19,000 volts to the convergence plates of the crt. The secondary of HCT-2 adds a horizontal parabolic waveform to the convergence-plate voltage for horizontal dynamic convergence. The high voltage for the ultor terminal of the crt is the sum of the voltage from high-voltage rectifier V801 and the voltage produced by X805.

One secondary winding on the flyback transformer has one lead grounded and the other lead connected to diode X604 through surge resistor R612. This rectifier circuit produces the 18 volts dc used by most of the circuits in the receiver. Capacitor C612 filters this voltage. A tap on the primary winding of the flyback transformer connects to a rectifier and filter circuit consisting of X605, R611, and C613. The 200-volt output of this network supplies the voltage for the color-video output amplifiers. By effectively connecting this rectifier circuit in series with the 115-volt dc supply, the portion of the 200 volts that must be supplied by the flyback transformer is substantially reduced.

Another secondary winding on the flyback transformer supplies ac to rectifier X606. The output voltage from this rectifier is applied to the horizontal-centering control (R603). The horizontal-deflection coils, the primary of the horizontal-output transformer, L601, and L602 form a series circuit connected from a tap on R603 to the movable arm. The amplitude and polarity of the current through the yoke coil can be varied by the position of the arm. This will shift the static position of the raster in a horizontal direction, thus effecting the centering of the picture.

OTHER SOLID-STATE HORIZONTAL-OUTPUT CIRCUITS

The solid-state horizontal-output and high-voltage circuit shown in Fig. 5-18 is used by General Electric. In contrast to the solid-state circuit just discussed, this arrangement somewhat resembles a tube-type horizontal-sweep circuit. Switching pulses for horizontal-output transistor Q71 are provided by a horizontal buffer stage. The horizontal-output transistor and damper diode D63 provide switching for the horizontal-yoke windings in a conventional manner. The collector voltage for Q71 is from a regulated 130-volt source in order to provide high-voltage regulation. The setting of the high-voltage control in the regulator circuit

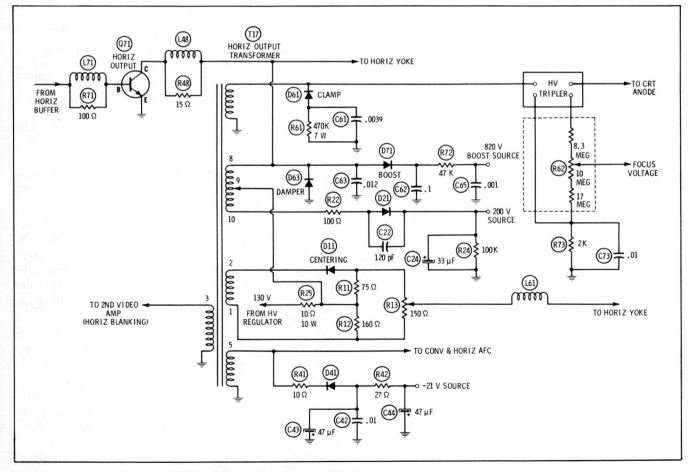

Fig. 5-18. Solid-state horizontal-output and high-voltage circuit used by General Electric.

determines the output voltage from the regulated source.

A secondary winding on the horizontal-output transformer provides a high-voltage pulse that is fed to the high-voltage tripler. A clamp circuit, consisting of D61, R61, and C61, clamps the negative portion of the high-voltage pulse during retrace. This clamp circuit improves high-voltage regulation when beam currents are low. The high-voltage tripler rectifies the high-voltage pulse and approximately triples its amplitude. With beam currents of 1.7 mA, the output from the tripler is 23.5 kV and the output is approximately 26.5 kV with minimum brightness and contrast. The focus voltage is developed across a focus divider connected to a tap on the high-voltage tripler. Focus divider R62 includes a 10-megohm potentiometer which is used as the focus control.

A horizontal centering network consisting of D11, R11, R12, and R13 is connected across terminals 1 and 2 of the horizontal-output transformer. The pulse voltage at terminal 2 is rectified by diode D11 and produces a small dc voltage that is negative at the high end of the horizontal centering control and positive at the low end. By adjusting the horizontal centering control, a small dc current can be made to flow through the horizontal-yoke windings. The direction and amplitude of this current will determine the horizontal position of the raster.

The pulse voltage at terminal 5 of the horizontal-output transformer is rectified by diode D41 to produce the −21-volt source for the i-f, low-level video, and RGB amplifier circuits. The pulse voltage at terminal 5 is also applied to the convergence and horizontal-afc circuits. The pulse voltage at terminal 10 is rectified by D21 to produce the +200-volt source that is supplied to the crt screen controls and other circuits. The 820-volt boost source is supplied by boost rectifier D71. The pulse at terminal 3 of the horizontal-output transformer is fed to the second video amplifier for horizontal blanking.

Circuits similar to the one shown in Fig. 5-18 are found in many late-model, solid-state color receivers. A major difference between these circuits and their tube-type counterparts is the effect of loss of the horizontal drive signal. Tube-type horizontal-output circuits will go into saturation when the drive signal is missing and the tube will quickly overheat. However, when there is no horizontal drive signal for the circuit in Fig. 5-18, the transistor will be cut off and collector current will cease.

LOW-VOLTAGE SUPPLY

The low-voltage supply in a color receiver is similar to the low-voltage supply used in a monochrome receiver in that it furnishes the B+ and filament voltages. Although these voltages are developed in a conventional manner, color receivers generally demand more power than the average monochrome receiver. This additional power requirement is due to the greater number of stages needed for color reproduction.

The low-voltage power supply for the RCA CTC16 chassis is shown in Fig. 5-19. The B+ voltage is developed across filter capacitor C1 by four silicon rectifiers connected in a bridge configuration. The output from the bridge is 410 volts at 480 milliamperes.

Heater voltages are supplied by two filament windings. One 6.3-volt winding, connected to the 260-volt source, provides filament voltage for the 6BK4 regulator tube. The other 6.3-volt winding, which is grounded at one end, supplies filament power to all the remaining tubes.

The power supply used in the General Electric color receiver chassis CA is shown in Fig. 5-20. The B+ voltage is developed in a slightly different manner in this receiver. The ac voltage from the secondary winding of the power transformer is applied to a voltage-doubler circuit, which utilizes two silicon rectifiers and a pair of 160-μF capacitors.

During a half cycle, current flow is from the positive plate of C3 through thermistor R1, through the transformer secondary winding, through rectifier X1, and back to the negative side of C3. This action places a potential of approximately 210 volts across C3. The voltage across the secondary of T1 is reversed during the next half of the cycle. This causes current to flow from the positive plate of C4 through rectifier X2, transformer T1, and thermistor R1, to the negative plate of C4. A voltage of approximately 210 volts appears across C4. By the use of large values for C3 and C4, the ripple content of their respective potentials is minimized. Since C3 and C4 are connected in series, the voltages appearing across them are aiding each other and the total output is 420 volts.

Notice that both of these supplies have automatic degaussing arrangements. These circuits are designed to degauss the picture tube each time the receiver is turned on.

The degaussing circuit is composed of three main elements: the degaussing coil or coils, a var-

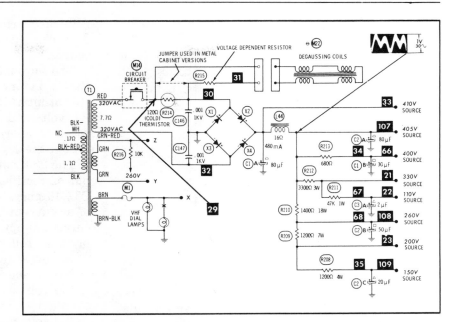

Fig. 5-19. Low-voltage power supply used in the RCA CTC16 chassis.

istor or voltage-dependent resistor, and a thermistor.

The coils are positioned about the perimeter of the picture-tube face, since it is the aperture mask that is to be degaussed (demagnetized).

The varistor is a voltage-dependent resistor that changes resistance with any change in applied voltage. A high voltage produces a low resistance.

The thermistor is a temperature-sensitive resistor. The resistance of the thermistor decreases as the temperature rises. In this application, the increase in temperature is provided by the flow of current through the unit.

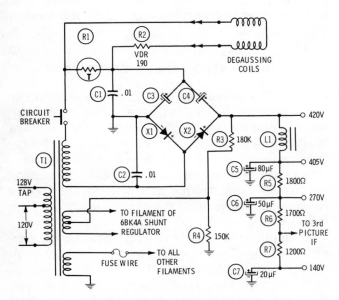

Fig. 5-20. Low-voltage power supply used in General Electric Chassis CA.

When the receiver is turned on, the thermistor has a resistance of approximately 120 ohms. The voltage drop across the thermistor is applied to the degaussing coils and the series varistor. Since the voltage applied to the varistor is relatively high, the resistance is low and most of the voltage appears across the degaussing coils. A high current flow in these coils produces an alternating magnetic field around and through the aperture mask. Current through the thermistor causes the resistance to decrease due to heating, and this lowers the voltage being applied to the coils and varistor. After a short interval, the voltage across the thermistor is only about 1 volt, and the resistance of the varistor is high enough to reduce the current flow in the coils to a negligible amount.

The RCA CTC40 chassis power supply provides four dc sources for general circuitry requirements and two ac power sources. The ac power sources are for the crt filaments and pilot lamps.

Power-supply switching circuits allow the CTC40 to take advantage of the "instant on" characteristics of solid-state devices. This switching circuitry is illustrated in Fig. 5-21.

The ac power is applied through the line filter and circuit breaker to the master power switch, S1. The master power switch applies power through the "instant pic" switch, S2, to both the dc supply transformer, T1, and the crt filament transformer, T2. However, when switch S2 is in the "off" position, reduced power is supplied to filament transformer T2 through a 680-ohm, 3-watt resistor, R1. Using this method, the crt filament is kept "warm" until full power is applied by

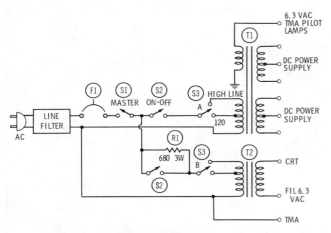

Fig. 5-21. Power-supply switching circuit used in RCA CTC40 chassis.

closing switch S2. This design ensures the full operation of the CTC40 within four to five seconds after turn on.

The master power switch, S1, is a rotary-type switch located at the top of the auxiliary user-controls bracket. Switch S2 is a push-pull switch located at the top of the user-controls panel and is adjacent to the brightness control.

The dc power supply provides four separate dc sources from three separate rectifier circuits. This is illustrated schematically in Fig. 5-22.

Rectifiers X1 through X4 are responsible for providing both the 82- and 30-volt sources. The 82-volt supply is derived from the full-wave bridge configuration of rectifiers X1 through X4. The transformer secondary winding that feeds this bridge circuit is center tapped and is used with rectifiers X2 and X3 to form a full-wave, center-tapped circuit with an output of 30 volts. A second

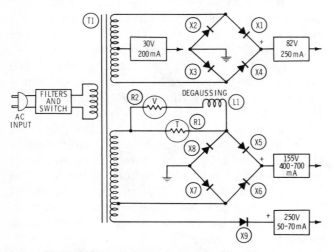

Fig. 5-22. Complete low-voltage power supply for RCA CTC40 chassis.

full-wave bridge circuit of rectifiers X5 through X8. The output of this circuit is the 155-volt source.

The automatic degaussing circuit is coupled to the secondary winding of T1 that feeds rectifiers X5 through X8. This circuit consists of thermistor R1, voltage-dependent resistor R2, and degaussing coil L1. Operation of this circuit is the same as that of the degaussing circuit previously described.

The 250-volt dc source is obtained from the output of rectifier X9. During normal operation, the CTC40 chassis draws approximately 1.8 amperes of ac current at 120 volts ac input. The average dc current supplied by each source of the power supply is as follows:

82-volt source—200 mA
30-volt source—250 mA
155-volt source—400-700 mA (varies with beam current)
250-volt source—50-70 mA (varies with beam current)

When cold, thermistor R1 has a relatively high resistance, permitting most of the ac current to flow through the voltage-dependent resistor R2 and the degaussing coil. This action creates a magnetic field about the coil, which degausses the picture tube. As the thermistor warms because of some small initial current through it, its resistance decreases and it passes on an increasingly larger share of current. As less current flows through the voltage-dependent resistor, its resistance increases, further restricting the current through the degaussing coil. Approximately five seconds after instrument turn-on, the current through L1 has decreased to zero, thus completing the degaussing action.

REGULATED POWER SUPPLIES

Since the operation of transistors and other semiconductor devices can be seriously affected by small voltage changes, the power-supply voltage must be relatively constant and ripple free. Therefore, power supplies used in solid-state color receivers are generally well regulated. These regulator circuits range from simple zener-diode circuits to complex active filters using as many as four transistors or an IC.

A simplified schematic of a regulated power supply used by Zenith is shown in Fig. 5-23. A unique feature of this circuit is the voltage-regulating transformer, T205. The primary winding of the transformer is loosely coupled to the secondary winding, which is in parallel with a 3.5-μF oil-

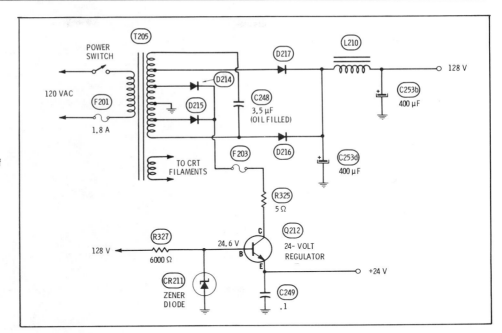

Fig. 5-23. Simplified schematic of regulated power supply used by Zenith.

filled capacitor, C248. The secondary winding and C248 form a resonant circuit that is tuned to the 60-Hz line frequency. With the secondary circuit resonating, the secondary voltage will increase to the point that the core material becomes saturated. Once saturation occurs, the secondary voltage will not increase any further and the output voltage will remain relatively constant. It should be noted that the output waveform for this transformer is more of a square wave than a sine wave.

The loose coupling between the primary and secondary windings makes the transformer action more dependent on core saturation. Another advantage of this loose coupling is that it will limit the short-circuit current. If a short circuit occurs, the current will increase to only approximately twice its normal value. A steady short circuit will trip the circuit breaker, and the transformer will not be damaged. The voltage-regulating transformer provides reasonable regulation for the 128-volt source with the power-line voltage ranging from 95 volts to 140 volts.

Additional regulation for the 24-volt source is provided by series regulator transistor Q212. The base voltage is fixed at 24.6 volts by zener diode CR211. The output voltage at the emitter will be the base voltage less the base-to-emitter forward voltage drop, or 24 volts.

Up to this point, the color-receiver sections that are comparable to those contained in monochrome receivers have been discussed. The next chapter will discuss the circuits that extract the chrominance signal from the composite color signal, and will also discuss the circuits used to demodulate the chrominance signal in preparation for its application to the picture tube.

QUESTIONS

1. What is the purpose of the delay line in a color television receiver?
2. What factor makes agc more important in a color receiver than in a monochrome receiver?
3. Does the horizontal-oscillator circuit in a color receiver differ from that in a monochrome receiver?
4. Why is it important to maintain a relatively constant ultor voltage on a color picture tube?
5. Describe the operation of an automatic degaussing circuit.

Chapter 6

Bandpass-Amplifier, Color-Sync, and Color-Killer Circuits

The material in the preceding chapters of this section described the color-receiver circuits that correspond very closely to those found in monochrome receivers. In the remaining chapters of this section, the circuits that deal with the proper reproduction of color are discussed. These circuits are represented by the unshaded blocks in Fig. 4-1 of Chapter 4.

In order to be utilized in the color receiver, the 3.58-MHz chrominance signal must first be separated from the composite color signal. An amplifier stage having a frequency-limiting filter network is used for this purpose. This stage is called the bandpass amplifier. The chrominance signal is fed from the bandpass amplifier to the demodulators where the color-difference signals are extracted from the 3.58-MHz signal. In order for the latter function to take place, a continuous-wave (cw) signal is required by each of the demodulators. These cw signals are generated and controlled by a section referred to as the color-sync section of the receiver. A burst amplifier, a keyer, a 3.58-MHz oscillator, and a control circuit are used in the color-sync section.

Many tube-type color receivers employ two demodulators. In this case, the color-sync section must provide two 3.58-MHz cw signals with different phases. Most solid-state receivers use three color-demodulator stages. When three demodulators are used, it is necessary to provide a 3.58-MHz signal with a different phase to each of the demodulators. The operation of demodulator circuits is discussed fully in the next chapter.

During the reception of a monochrome signal by the color receiver, a means of cutting off the chrominance channel is provided. This function is performed by the color-killer section which auto-

matically disables the chrominance channel when no color signal is being received.

These three sections of the receiver—the bandpass amplifier, the color-sync section, and the color killer—are discussed in this chapter. First, the chrominance signal will be traced to the input of the demodulators. Then the manner in which the 3.58-MHz cw signals are developed by the color-sync section will be discussed. The color killer is controlled by the color-sync section, and it affects the operation of the chrominance channel.

BANDPASS AMPLIFIER

The purpose of the bandpass amplifier is to separate the chrominance signal from the composite color signal and feed it to the demodulators. The input signal for a typical bandpas-amplifier section is the composite color signal, which includes the chrominance, luminance, burst, synchronizing, and blanking signals. The takeoff point for the chrominance signal is usually in the first video amplifier stage.

Only the chrominance portion of the composite color signal appears at the output of the bandpass-amplifier circuit. Between the signal takeoff point and the input of the demodulators, any remaining 4.5-MHz sound signal has been attenuated, the luminance signal has been blocked, and the color-burst and synchronizing signals have been keyed out.

One type of circuit used to separate the chrominance signal from the composite color signal is shown in Fig. 6-1. This circuit employs one stage of amplification and is referred to as the chroma-bandpass amplifier. The name specifies that this stage passes and amplifies the chrominance por-

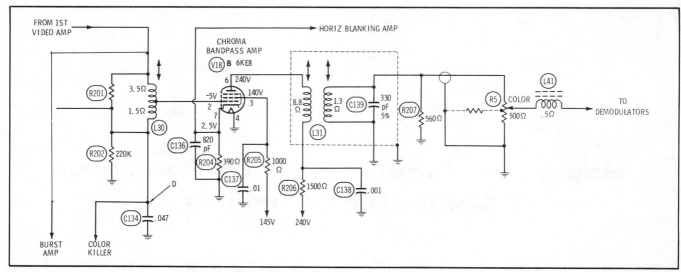

Fig. 6-1. Chroma-bandpass amplifier in a tube-type color-television receiver.

tion of the video band of frequencies. Let us see how this is accomplished.

The composite color signal is taken off at the plate of the first video amplifier. The NTSC color-bar generator signal is shown by the waveform in Fig. 6-2A. Fig. 6-2B shows the waveform produced by a keyed-rainbow generator. The white, bar-shaped portions of the waveform in Fig. 6-2A are produced by the chrominance signal. Because of the presence of the luminance signal, the dc level of the composite signal varies according to the brightness of the colors represented. Notice also that the horizontal-sync pulse is quite prominent in this waveform. Since all of the bars in the keyed-rainbow pattern are the same brightness, there is no variation in the dc level of its composite signal.

In Fig. 6-3, the waveform at the grid of the chroma-bandpass amplifier contains only the 3.58-MHz chroma signal, including the burst. The chroma-takeoff coil in conjunction with the associated circuitry is tuned to a bandpass of approximately 1 MHz. This circuit rejects all signals except the 3.58-MHz chrominance signal. Notice that the amplitude excursions and the prominent

sync pulse are not present at the grid of the bandpass amplifier and that the dc level is the same throughout the waveform.

The burst shown in Fig. 6-3 is not required at the input to the demodulators. The keying pulse shown in Fig. 6-4, which is obtained from the cathode of the blanker tube, drives the cathode of the chroma-bandpass amplifier positive during the horizontal-retrace interval and prevents the burst from reaching the color demodulators. The result of keying the chroma-bandpass amplifier is shown in Fig. 6-5. The signal is shown at the output of the chroma-bandpass amplifier. Note that the burst has been eliminated from the color signal.

The information contained in the signal of Fig. 6-5 is only the 3.58-MHz chrominance signal necessary to reproduce the color portion of the television picture. The sync pulses, luminance signals, blanking pulses, and the color burst have been removed. The frequencies passed on to the color demodulators are limited to the 0.5-MHz sidebands of the 3.58-MHz chrominance subcarrier. A frequency-response curve is shown in Fig. 6-6, indicating the frequency limits for the chroma-bandpass amplifier in Fig. 6-1.

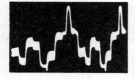

(A) NTSC
color-bar generator.

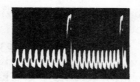

(B) Keyed-
rainbow generator.

Fig. 6-2. Signal from the plate of the first video amplifier in Fig. 6-1.

(A) NTSC
color-bar generator.

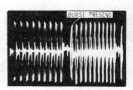

(B) Keyed-
rainbow generator.

Fig. 6-3. Signal at the grid of the chroma bandpass amplifier in Fig. 6-1.

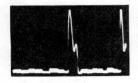

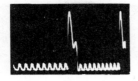

(A) NTSC
color-bar generator.

(B) Keyed-
rainbow generator.

Fig. 6-4. Keying pulse at the cathode of the chroma-bandpass amplifier in Fig. 6-1.

A two-stage chroma-bandpass amplifier is shown in Fig. 6-7. This circuit performs the same function as the one just discussed; however, the signal is processed in a slightly different manner. The input signal from the video cathode follower, which also acts as a chroma amplifier, contains both chrominance and luminance information. The luminance portion of the signal is prevented from reaching the grid by the high impedance of the 3.3-pF capacitor C116.

(A) NTSC
color-bar generator.

(B) Keyed-
rainbow generator.

Fig. 6-5. Signal at the output of the chroma-bandpass amplifier in Fig. 6-1.

The amplified chrominance signal applied to the grid of V13A is shown in Fig. 6-8. Bias is also applied to this grid through the color-level control from the color killer. With a color signal present, the bias is approximately 0.4 volt. During reception of a black-and-white signal, the color killer places about 20 volts bias on the grid, cutting off the second chroma-bandpass amplifier.

The output of the second bandpass amplifier is separated into two identical signals. This signal is shown in Fig. 6-9. One signal drives the X demodulator grid, and a second signal drives the Z demodulator grid.

Note that in this circuit the 3.58-MHz burst is not eliminated. A horizontal-blanking pulse is applied to the cathodes of the G − Y, B − Y, and R − Y amplifiers. The picture-tube screen is blanked out during the horizontal-retrace time, and the 3.58-MHz burst is effectively eliminated.

Shown in Fig. 6-10 is a version of bandpass amplifiers employed in a Zenith hybrid chassis. The foregoing discussion of theory of operation will also suffice for this chassis. The only signifi-

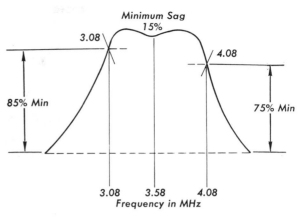

Fig. 6-6. Frequency-response curve of chroma-bandpass amplifier in Fig. 6-1.

cant change is the addition of the transistor, with the necessary bias and supply voltage circuitry.

CHROMA-BANDPASS AMPLIFIER (SOLID-STATE CIRCUITS)

In the circuit shown in Fig. 6-11, the composite video signal from the first video-amplifier emitter is coupled to the first chroma amplifier through a coupling network consisting of capacitor C1 and inductor L1. The values of these components are selected to provide attenuation of low-frequency video signals and to pass frequencies in the chroma i-f band. The amount of amplification of these signals by the first chroma i-f amplifier (Q1) is dependent on the amplitude of the acc bias applied to the base of Q1 through resistor R1. The output of the first chroma i-f amplifier is then coupled through C2 to the base of the second chroma i-f amplifier, Q2.

The output of the second chroma i-f amplifier is applied across a resonant circuit consisting of bandpass transformer T1 and capacitor C3. The resistor (R2) shunting the resonant circuit ensures the proper bandpass for the circuit. The bandpass transformer has a top and bottom slug adjustment. The top slug is adjusted for 3.1 MHz; the bottom slug for 4.1 MHz. Proper adjustment of both slugs allows the bandpass transformer to pass the full range of chroma i-f frequencies.

The output of the resonant circuit is then fed to the tint and color controls through a coupling network consisting of capacitor C4 and resistor R3. The tint control functions to provide either capacitance or inductance loading of the bandpass transformer. This provides a phase shift of the entire chroma i-f signal and results in full tint-control range. The color control functions to vary the am-

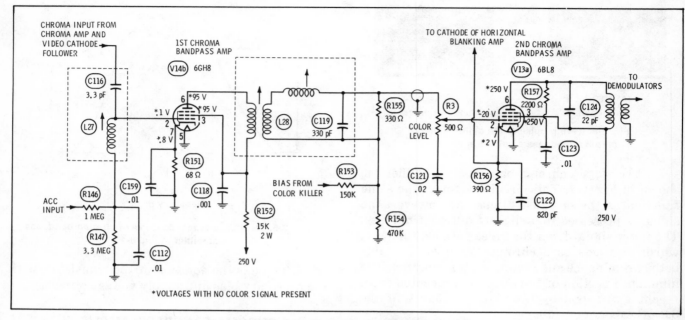

Fig. 6-7. Two-stage chroma bandpass amplifier used by Philco.

plitude of the color signals applied to the base of the chroma output stage, Q3.

The base of Q3 receives bias from the collector of the color-killer transistor. A blanking pulse from the emitter of the blanker transistor is coupled through inductor L2 to the emitter of Q3. The blanking pulse effectively eliminates the 3.58-MHz burst signal at the chroma output collector, thus permitting only true color signals to be coupled through transformer T2 to the demodulator circuits, and providing suitable dc reference for the chroma signal.

COLOR SYNCHRONIZATION

In order to properly reproduce the colors of a televised image, the modulation of the chrominance subcarrier at the transmitter must be reversed at the receiver. It may be recalled that the modulation process at the transmitter involves the use of a 3.58-MHz subcarrier. This subcarrier is applied in quadrature to two doubly balanced modulator circuits. Simultaneously, the I signal is applied to one balanced modulator, and the Q signal is applied to the other. The 3.58-MHz subcarrier is cancelled, and the resultant output is the chrominance signal. This is a 3.58-MHz signal which varies in amplitude and phase. It is added to the composite video signal at the transmitter.

Recovery of the color-difference signals in the receiver is accomplished by reversing the modulation process. This requires a locally generated 3.58-MHz cw reference signal that is applied to two or three demodulator circuits. Accuracy in the demodulation process is attained by regulating this reference signal so that a definite phase relationship with the subcarrier is maintained for each of the demodulators. It is the function of the color-sync section to generate the local 3.58-MHz reference signal and regulate its frequency and phase.

A block diagram of a color-sync section in a tube-type color receiver is shown in Fig. 6-12. An oscillator generates a 3.58-MHz sine wave that is

(A) NTSC
color-bar generator.

(B) Keyed-
rainbow generator.

Fig. 6-8. Signal at the grid of the second chroma bandpass amplifier in Fig. 6-7.

(A) NTSC
color-bar generator.

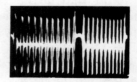

(B) Keyed-
rainbow generator.

Fig. 6-9. Signal at the output of the second chroma bandpass amplifier in Fig. 6-7.

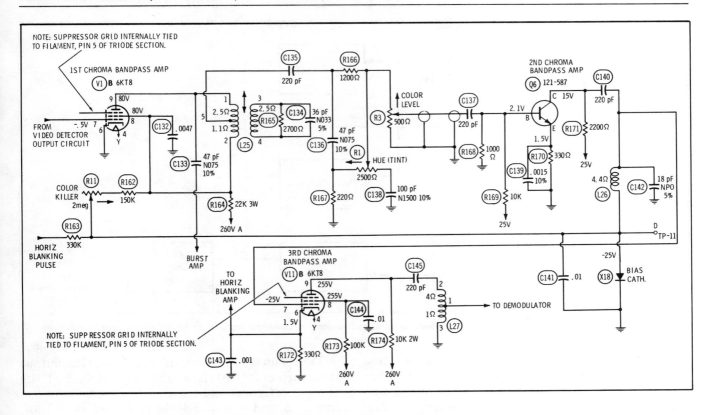

Fig. 6-10. Bandpass amplifier used in Zenith hybrid color chassis.

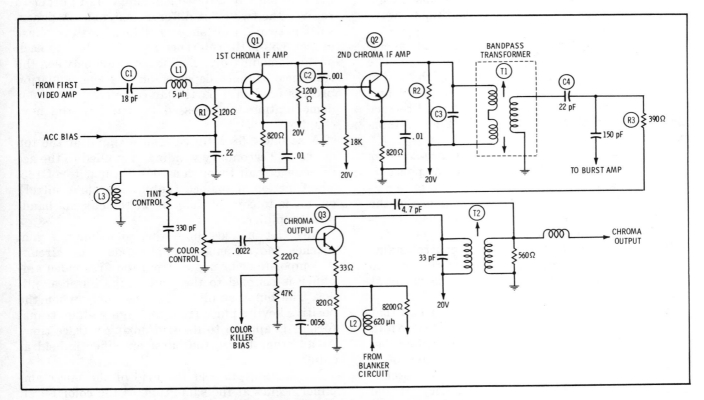

Fig. 6-11. Solid-state chroma-bandpass amplifier.

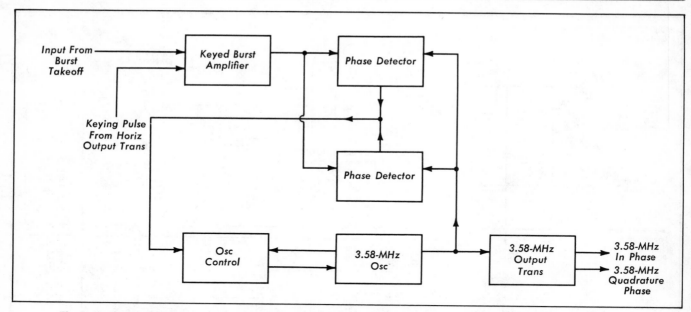

Fig. 6-12. Block diagram of a tube-type color-sync circuit using phase detectors and an oscillator control stage.

amplified by the following stage. The output circuit of the amplifier stage provides two 3.58-MHz signals having a phase difference determined by the demodulation process used. These are the cw reference signals used in the demodulation process. The proper frequency and phase of the 3.58-MHz oscillator signal is assured by a phase comparison between the in-phase continuous-wave component and the color burst.

A composite color signal from the burst-takeoff point is applied to the burst-amplifier stage. The operation of this stage is controlled by the action of the keying pulse which permits the burst amplifier to conduct only during horizontal-retrace time. As a result, the output of this stage contains only the color burst.

The color burst is applied to the phase-detector stages (usually two diodes) where its phase is compared with the phase of the cw reference signals. Any error in the frequency or phase of the reference signal that is used for comparison is detected by the phase detectors, and a correction voltage is applied to the oscillator control stage. This correction voltage causes the control stage to affect the oscillator circuit until it is operating at the proper frequency and phase.

A schematic diagram of a color-sync circuit that operates in the manner just described is shown in Fig. 6-13. This circuit is used in a late-model tube-type color receiver. The pentode section of a 6GH8A (V22) is used as the chroma-reference oscillator and determines the frequency of the 3.58-MHz reference signal.

The primary winding of the 3.58-MHz output transformer (L39) is in the plate circuit of the chroma-reference oscillator. This transformer is tuned so that the 3.58-MHz reference signal appears across its secondary winding. The 3.58-MHz cw signal at the center tap of the secondary winding is applied to the two demodulators in order to recover the two color-difference signals. A phase-shift network (not shown) determines the phase of the 3.58-MHz reference signal applied to each of the demodulators. The phase angle between the reference signals depends on the demodulation process used in a particular color receiver. Color demodulation is discussed in detail in the next chapter.

A second 3.58-MHz reference signal at the top of the L39 secondary winding is applied to the acc detector circuit through a 100-pF capacitor, C165. The function of the acc (automatic color control) circuit is to control the gain of the chroma band-pass amplifier.

Now let us see how the color-burst signal reaches the chroma-sync phase-detector circuit. A composite color signal from the first video amplifier is coupled to the grid of the burst amplifier. The burst amplifier conducts only when the positive keying pulse from the high-voltage transformer is applied to its grid during retrace time. At all other times, the burst amplifier is held at cutoff.

The keying pulse at the grid of the burst amplifier occurs at the same time as the color burst. Since the burst amplifier conducts only during the

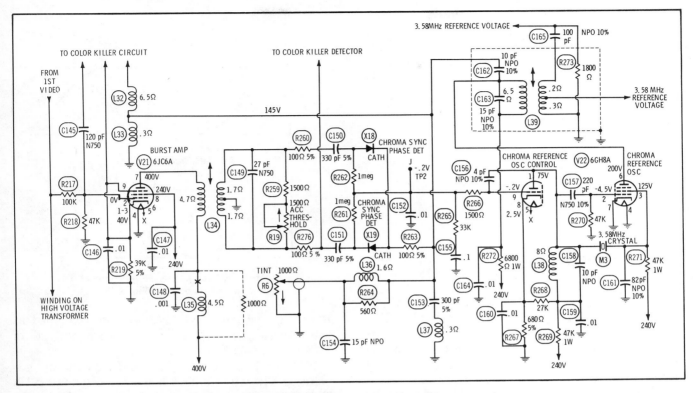

Fig. 6-13. Color-sync circuit used in a tube-type color receiver.

period of the positive keying pulse, only the color burst is amplified. The color burst appears across the primary winding of the burst transformer, L34. The secondary winding of L34 is tightly coupled to the primary. Resistor R259 in series with control R19 across the secondary winding lowers the Q factor of this winding to obtain a broader frequency response. This enables the burst transformer to be tuned so that the complete burst signal occurs at the center of the response curve.

The signal at one end of the secondary winding of L34 is 180 degrees out of phase with the signal at the other end. The two ends of the secondary winding are coupled to the phase-detector diodes by C150 and C151. Thus, the signal at the anode of diode X18 is 180 degrees out of phase with the

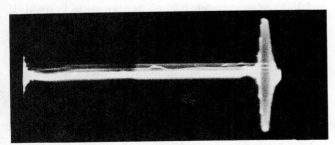

Fig. 6-14. Burst signal at chroma-sync phase detector in Fig. 6-13.

signal at the cathode of diode X19. Fig. 6-14 shows the color burst signal observed at either diode. The 180-degree phase difference is not evident on an oscilloscope.

If accurate demodulation is to be obtained, the phase of the I reference signal must lag the phase of the color burst by 57 degrees. It will be shown later that the action of the detector circuit causes the phase of the I reference signal to lead the phase of the color burst at the anode of X18 and to lag the phase of the color burst at the cathode of X19 by 90 degrees.

When L34 is properly tuned, no phase change takes place during the energy transfer between the windings of this transformer; therefore, the color burst at the plate of V21 will have the same phase as the signal at the anode of X18. The polarity of the color burst is inverted through the burst amplifier; therefore, the signal at the grid of this tube will be 180 degrees out of phase with the signal that appears at the plate. As a result, the color burst at the grid of V21 will have the same phase as the signal at the cathode of X18.

Let us assume that the vector at zero degrees (in Fig. 6-15) represents the phase angle of the color burst at its takeoff point at a particular time. The desired phase angle of the I reference signal at this time is represented by the vector shown at

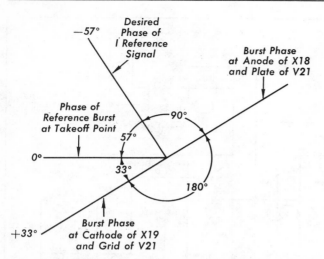

Fig. 6-15. Vector diagram showing the phase angles of the I reference and burst signals in the circuit of Fig. 6-13.

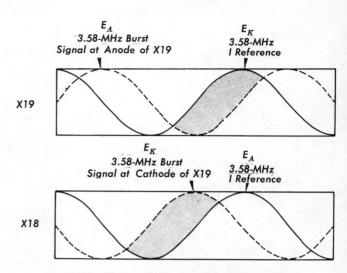

Fig. 6-16. Normal phase relationships at the phase detector in Fig. 6-13.

−57 degrees. Since the burst phase at the grid of V21 leads the phase of the I reference signal by 90 degrees, the phase of the color burst at this point can be represented by the vector shown at +33 degrees. This means that the phase angle of the burst signal must be advanced by 33 degrees between the takeoff point and the grid of the burst amplifier if the local 3.58-MHz oscillator signal is to have the correct phase angle.

Refer to Fig. 6-13 and note that the tint control in conjunction with L36 and C154 forms a variable phase-shift circuit for the chroma reference signal. The tint control functions as a vernier, and its setting will determine the phase of the chroma reference signal applied to the phase-detector diodes.

When the 3.58-MHz oscillator signal has the correct frequency and phase, the I reference signal will lead the color burst at the anode of X18 and lag it at the cathode of X19 by 90 degrees. This phase relationship is shown in Fig. 6-16. Each of the diodes will conduct when the anode voltage E_A exceeds the cathode voltage E_K. The period of conduction for each diode is indicated by the shaded areas.

During the time when diode X18 conducts, electrons flow downward through resistor R262. When diode X19 conducts, electrons flow upward through resistor R261. If the phase relationships between the color burst and the I reference signal are like those shown in Fig. 6-16, the current through the two diodes will be equal. As a result, the voltage at the junction of R261 and R262 will be zero. If the current through X18 is greater than the current through X19, a negative dc voltage will be developed at the junction of R261 and

R262. A positive dc voltage will be developed at this junction if X19 conducts more heavily than X18.

If the phase of the 3.58-MHz oscillator advances, X19 will conduct more heavily than X18. This will cause a positive voltage to be developed at the junction of R261 and R262. This voltage is of the proper polarity to cause the phase of the oscillator signal to be retarded until its phase is correct. If the oscillator should slow down, X18 will conduct more than X19. A negative voltage will be applied to the grid of the oscillator control tube (triode sction of V22) through resistor R266. This will cause the phase of the oscillator to advance.

Another way of explaining how the dc correction voltage is developed is to consider the voltage developed across C152. If X19 conducts more heavily than X18, more electrons will flow from the top plate of C152 than will flow into it and a positive dc correction voltage will exist. If X18 conducts more than X19, more electrons will flow into the top plate of C152 than will flow out of it and a negative correction voltage will be developed. Regardless of which method of explanation is used, it can be seen that a dc voltage of the proper polarity will be developed at TP2.

The grid of the oscillator control tube has no dc path to ground, and there is no dc current through R266. This means that there will be no dc voltage dropped across these components, and any dc voltage developed at the junction of R261 and R262 will appear at the grid of the control tube.

The bias on this stage is not easily affected by noise pulses. Any noise pulses occurring at retrace time would be amplified by the burst amplifier. The polarity of the resultant voltage could be positive or negative, depending upon the polarity of the noise pulse. C152 responds to this pulse and accordingly charges. C152 will discharge through R265 and C155. The discharge current tends to charge C155; however, the large value of C155 and the current-limiting action of R265 prevent any appreciable change in the voltage on C152 due to noise. In this way, the network composed of R265 and C155 forms a sort of noise-immunity circuit.

The schematic in Fig. 6-17 shows the equivalent circuit of the control stage at 3.58 MHz. Consider for the moment that the 3.58-MHz signal from the generator is not yet applied across the tube and that its frequency and phase are as shown by Fig. 6-18A. If this signal is then applied across the tube, the 4-pF capacitor C156 provides a coupling of high capacitive reactance from plate to grid. In a capacitive circuit, the current leads the voltage by 90 degrees; therefore, the charge and discharge current of C156 leads the voltage across this capacitor by 90 degrees. Since the same current flows through R266, the current through this resistor leads the voltage that is applied from the generator by approximately 90 degrees. The voltage developed across R266 is in phase with this current; therefore, it leads the applied voltage by 90 degrees. This is the signal voltage at the grid of the tube and is represented by Fig. 6-18B.

The signal at the grid of the tube is amplified, and the polarity is inverted. The signal output of the tube is represented by the drawing in Fig. 6-18C. It can be seen that signal C lags signal A by 90 degrees. Actually, signals A and C do not exist as shown, since the signal at the plate of the control tube is a combination of signals A and C. The phase of this resultant signal is somewhere between the phase of the generator signal and the tube signal, favoring the phase of the signal having the greater amplitude. The signal supplied by

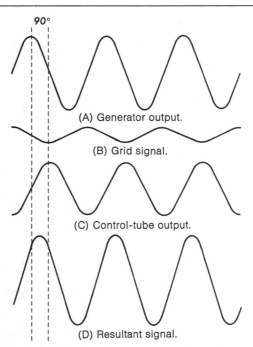

Fig. 6-18. Signals present in the control-tube circuit. The resultant signal is developed by the combination of signals A and C.

the generator has a constant amplitude; whereas, the amplitude of the signal supplied by the tube will vary with changes in the bias of the stage.

The signal at the plate of the tube, and formerly represented as signal A, has assumed a phase represented by signal D. Note that signal D lags signal A by almost 45 degrees; therefore, the phase of the applied signal has changed. It can therefore be seen that the reactance tube is effective in changing the phase of the 3.58-MHz signal.

Let us consider the operation of the complete circuit shown in Fig. 6-13. The oscillator begins to operate at a frequency which is dependent upon the characteristics of the crystal and upon the capacitive reactance contributed by the reactance tube. A sine wave is obtained from the tuned plate circuit of the oscillator. A sample of the I reference signal is fed through C162 to the phase detectors where its phase is compared with the phase of the color burst.

Let us assume for the moment that the phase of the oscillator signal has been retarded. This phase delay will be reflected in the phase of the I reference signal, and the voltages at the detector diodes will assume the phase differences shown in Fig. 6-19. It can be seen that X18 will conduct more heavily than X19. As previously discussed, a negative dc difference voltage is produced, and the bias on the control stage increases. The gain of the control tube is reduced, thereby reducing the capaci-

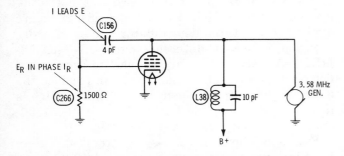

Fig. 6-17. Simplified version of oscillator-control circuit in Fig. 6-13.

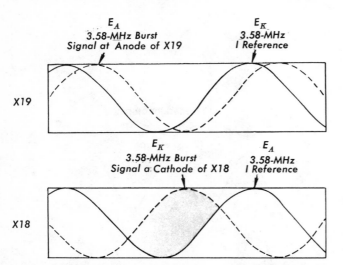

Fig. 6-19. Phase relationships between the signals applied to the phase-detector circuit in Fig. 6-13 when the phase of the 3.58-MHz reference oscillator is lagging.

tance of the reactance tube. This will allow the phase of the oscillator signal to advance. The bias on the control tube will steadily decrease until the phase of the oscillator is in step with that of the color burst.

If the phase of the oscillator signal should advance, the condition shown by Fig. 6-20 will exist at the detector diodes. A positive dc correction voltage will be developed across C152, because X19 will conduct more heavily than X18. The gain of the control stage will be increased, and the phase of the oscillator signal will be retarded. The conduction of X18 will slowly increase, and the conduction of X19 will slowly decrease. The positive correction voltage will gradually decrease as the

oscillator signal approaches the proper phase.

The control stage always has some effect upon the frequency of the oscillator signal because the capacitance contributed by the control stage is effectively connected in parallel with the crystal. A short time after the circuit is put into operation, a steady condition exists. The phase-detector circuit constantly supplies the necessary correction voltage to lock the oscillator signal into synchronization with the color burst.

SOLID-STATE COLOR-SYNC CIRCUITS

A color-sync circuit using transistors is shown in Fig. 6-21. The operation of this circuit is somewhat different than the tube-type circuit just discussed. A color signal from the output of the 1st chroma-bandpass amplifier is applied to the base of the burst amplifier Q306. At the same time, a horizontal-sync pulse from the sync separator is applied to the burst amplifier. This pulse turns on the burst amplifier during retrace time, allowing only the color burst to appear at the output. The sub-hue control (a service adjustment) in conjunction with L304, adjusts the phase of the color-sync signal applied to the base of Q306.

The color burst is amplified by Q306 and is coupled to the secondary winding of the burst transformer, T304. The burst signal is then coupled to ringing crystal X301, causing its output to be in phase with the incoming burst. The 3.58-MHz output from the crystal is then fed to Q307 where it is amplified. The 3.58-MHz signal at the collector of Q307 goes to the color-killer circuit and through the hue-control circuit to a second 3.58-MHz amplifier, Q308. The hue control varies the phase of the 3.58-MHz signal applied to the base of Q308. The output from the second 3.58-MHz amplifier is then applied to the 3.58-MHz oscillator, Q309. Positive feedback for the oscillator is furnished by a 39-pF capacitor, C348. The two 3.58-MHz reference signals developed at the top and bottom of the 3.58-MHz output transformer are 180 degrees out of phase. The two subcarrier signals are applied to the three demodulator stages in order to produce the color-difference signals. These 3.58-MHz reference signals are coupled to each set of demodulators in such a manner as to produce the desired color output. When the AUTO-AFT switch is in the on position, the hue is fixed by the 1500-ohm resistor (R342) across coil L308.

Fig. 6-22 shows a color-sync circuit using an IC. Since most of the signal-processing circuitry is contained within the IC, it will be more practi-

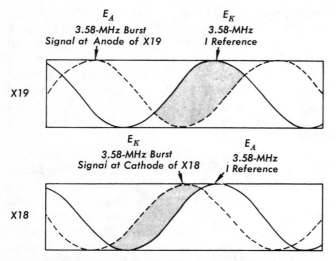

Fig. 6-20. Phase relationships between the signals applied to the phase detector circuit in Fig. 6-13 when the phase of the 3.58-MHz reference oscillator is leading.

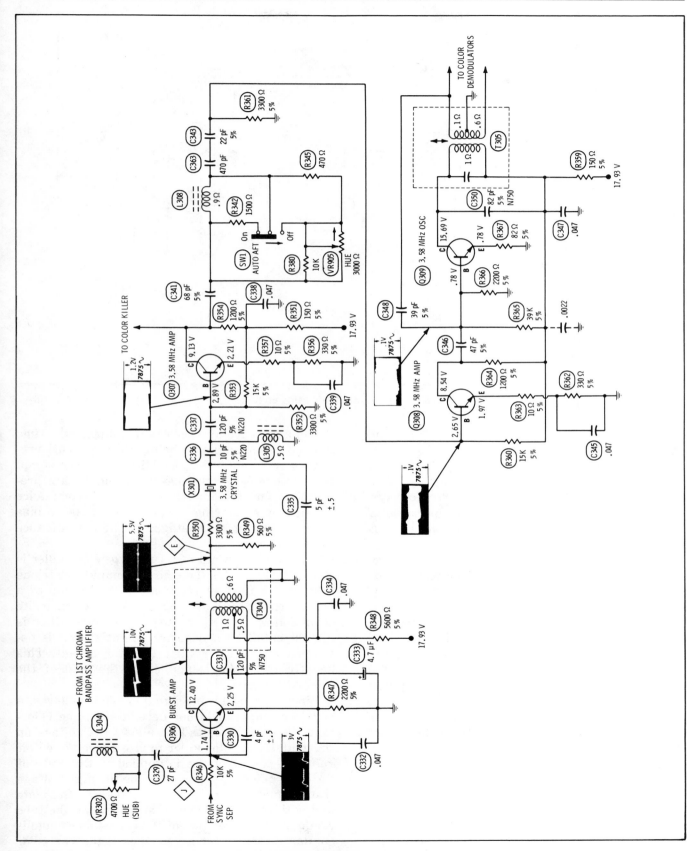

Fig. 6-21. A solid-state color-sync circuit.

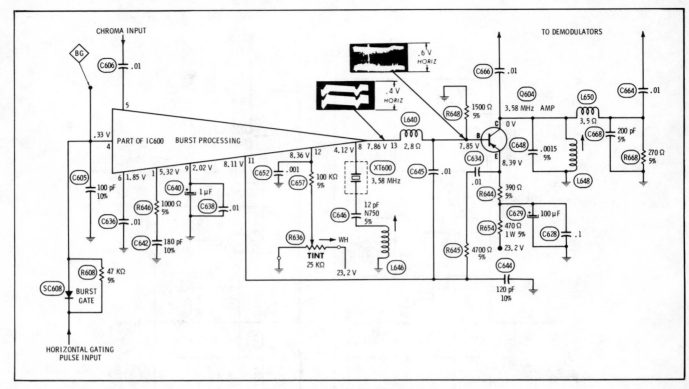

Fig. 6-22. Color-sync circuit using an IC.

cal to discuss this circuit in terms of the input and output signals. The chroma input signal is fed to pin 5 of IC600 through capacitor C606. Much of the color-processing circuitry, including the bandpass amplifiers, is contained in IC600. However, since we are concerned only with the operation of the color-sync circuit here, not all of the IC pins have been shown.

A horizontal gating pulse is fed to pin 4 to provide gating for the burst-processing circuit during horizontal retrace time. The burst signal is then passed to the internal 3.58-MHz subcarrier oscillator. The external frequency-determining components for the subcarrier oscillator include the 3.58-MHz crystal (XT600) and coil L646. The subcarrier output from pin 13 of IC600 is fed through L640 to the base of 3.58-MHz amplifier Q604. Two amplified 3.58-MHz subcarrier signals are then applied to the demodulators. One of these signals is taken directly from the collector of Q604, and the other signal passes through a phase-shift network consisting of L650, C608, and R668 to establish the proper demodulation axes.

COLOR KILLER

The purpose of the color killer in a color receiver is to prevent any signal from getting through the chrominance channel during the time a monochrome signal is being received. This prevents any signal other than the luminance signal from reaching the picture tube. Signals are prevented from passing through the chrominance channel by employing a color-killer stage to bias to cutoff one or more stages in the chrominance channel.

A simplified circuit of a tube-type color killer is shown in Fig. 6-23. This circuit employs a triode tube, which in most receivers is one section of a multipurpose tube. During the time a composite color signal is being received, the color killer is held at cutoff by a negative potential that is developed in the color-killer detector circuit. This negative potential is applied to the grid of the color-killer tube.

When a monochrome signal is being received, a negative potential is not developed in the color-killer detector circuit. This allows the voltage on the grid of the color killer to rise above the cutoff value. A positive pulse is applied to the plate of the color killer from a winding on the high-voltage transformer. With a positive voltage on the plate and with the grid above cutoff potential, the tube conducts. The operation of the color-killer stage is similar to that of a keyed agc stage. The color killer conducts during the time of a positive pulse

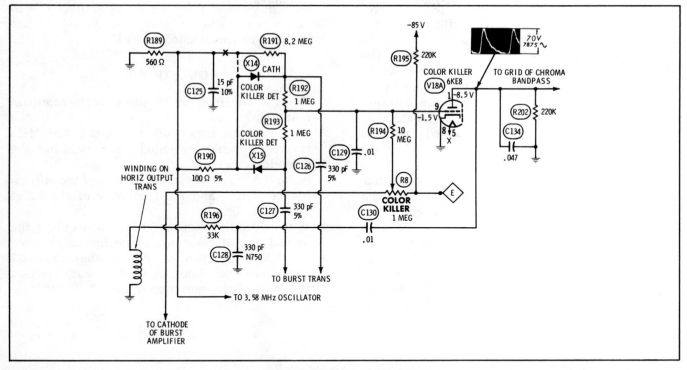

Fig. 6-23. Tube-type color-killer circuit.

from the transformer winding, and plate current charges capacitor C130. Between pulses this capacitor discharges through R202 and maintains a charge on capacitor C134. The negative voltage across C134 is sufficient to cut off the chroma-bandpass amplifier. Capacitor C134 acts as a signal bypass to ground for the tuned circuit in the grid of the chroma-bandpass amplifier and also acts as a filter for the horizontal pulses appearing on the plate of the color killer.

A color-killer threshold control is provided to permit adjusting the level at which the color killer begins conduction. This ensures a proper turn-on and turn-off of the chroma-bandpass amplifier.

A solid-state color-killer circuit is shown in Fig. 6-24. The control voltage for the color-killer circuit is derived from the acc detector. The output from the acc detector is applied to the base of the killer amplifier, Q706. When a color signal is being received, a negative voltage from the acc detector

Fig. 6-24. A solid-state color-killer circuit.

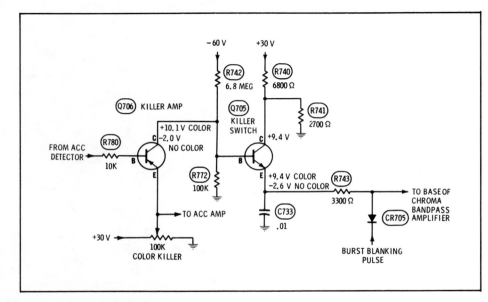

causes Q706 to conduct. This develops a positive voltage on the collector of the killer amplifier which is connected to the base of the killer switch, Q705. This positive voltage on the base of the killer switch turns it on, causing a positive voltage to be developed at the emitter. This positive voltage, in turn, is fed to the base of the chroma-bandpass amplifier, causing it to conduct. This allows the chroma signal to be processed normally.

When a black-and-white signal is being received, a positive voltage is produced by the acc detector. This positive voltage cuts off the killer amplifier and removes the forward bias from the killer switch. With killer switch Q705 cut off, a negative voltage will be developed at its emitter due to rectification of the burst blanking pulse by CR705. This negative voltage cuts off the chroma-bandpass amplifier, preventing any color "noise" from reaching the color demodulators. The setting of the color-killer control establishes the emitter bias for the killer amplifier.

Most late-model color receivers employ ICs in the chroma circuits. In these sets, the color-killer function is often included in an IC.

QUESTIONS

1. What is the purpose of the chroma-bandpass amplifier?
2. How are the sync pulses and the 3.58-MHz reference burst prevented from reaching the demodulators?
3. State the frequency in megahertz of the chrominance signal, and the frequency of the burst reference signal.
4. What circuit is responsible for preventing the appearance of color in a monochrome picture?
5. What is the purpose of the color-burst signal?
6. How does the "hue" or "tint" control affect the color of the reproduced scene?

Chapter 7

Color Demodulation

The function of the chrominance demodulators in a color receiver is the reverse of that of the modulators in a color transmitter. It may be recalled that the I and Q signals are used at the transmitter to modulate the amplitude of separate portions of a divided carrier. These portions have the same frequency, but they differ in phase by 90 degrees. The two portions of the carrier are suppressed, and the sidebands are combined to form the chrominance signal.

The demodulator circuits in a color receiver must be capable of detecting both the amplitude and the phase of the chrominance signal rather than just the amplitude. This is necessary because the amplitude of the subcarrier at the transmitter was modulated at zero phase by the I signal and was modulated at quadrature phase by the Q signal. The two modulated signals were then combined to form the chrominance signal. Basically, then, the chrominance signal consists of two sine waves in quadrature, and the amplitude of each is modulated. This means that the chrominance signal received by the demodulator circuits in a color receiver is varying in both amplitude and phase.

In order to reproduce the original colors, the demodulators must convert the chrominance signal into the I and Q signals. The three original signals for red, green, and blue can also be reproduced by recovering color-difference signals other than the I and Q signals. This will be discussed in the latter part of this chapter.

The type of chrominance demodulation most commonly used is synchronous demodulation, a process by which amplitude variations of a single phase of a carrier that is modulated at two different phases can be detected. A synchronous demodulator can be compared to the mixer stage in a superheterodyne radio receiver. The main difference is that the synchronous demodulator uses a locally generated reference signal which is of the same frequency as that of the color subcarrier, whereas a superheterodyne mixer stage requires a locally generated reference signal which has a higher or lower frequency than that of the incoming carrier.

A SIMPLIFIED DEMODULATOR CIRCUIT

A multigrid tube is often employed as a synchronous demodulator in tube-type color receivers. A simplified circuit of a synchronous demodulator is shown in Fig. 7-1A. At control grid No. 1 there are two sine waves arriving in quadrature (90-degree phase difference). Sine wave A is leading sine wave B by 90 degrees. At the time that sine wave A is at maximum, sine wave wave B is zero. It can be assumed that these two sine waves are the two modulated signals which are combined to form the chrominance signal.

Two supply voltages for grid No. 3 are available in this simplified circuit. One potential is zero, and the other is negative. When the voltage on grid No. 3 is zero, the tube will conduct. When grid No. 3 is switched to the negative potential, current will cease to flow through the tube.

Let us assume that the amplitude of sine wave A is to be detected by the synchronous demodulator. The grid No. 3 voltage is negative all the time except during time 2. At that time, the grid No. 3 potential is switched to zero volts, which means that current will flow through the tube. Notice that sine wave A reaches its maximum value at the same time. Each time that this sine wave reaches its positive peak, the tube will conduct because the grid No. 3 voltage is zero at the

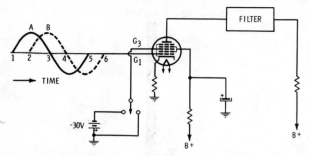

(A) Simplified synchronous-demodulator circuit.

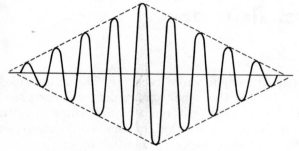

(B) Amplitude-modulated carrier.

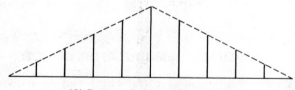

(C) Pattern of plate-current pulses.

Fig. 7-1. Illustration showing the result of synchronous demodulation.

same time. The amount of current that will flow is dependent on the value of the voltage applied to the control grid during time 2.

Let us now consider the manner in which the stage operates when sine wave B is fed to grid No. 1. The switching of grid No. 3 will be the same as in the previous case; that is, grid No. 3 will be switched to zero potential during time 2 and to a minus voltage the remainder of the time. This means that the stage can conduct during time 2 only. The amount of current is dependent on the amplitude of the voltage applied to grid No. 1, which in this case is zero. The amplitude of the plate-current pulses will be less when sine wave B is applied to the stage than when sine wave A is applied. The amplitudes of sine waves A and B are equal, and the two signals differ only in phase. If sine wave A were increased in amplitude, the resultant plate-current pulses would also increase. The plate-current pulses resulting from the application of sine wave B would remain the same even though the amplitude of B might change because this signal always passes through zero during the

time the tube conducts. This is the basis upon which a synchronous demodulator operates.

Fig. 7-1B shows an amplitude-modulated carrier. Fig. 7-1C illustrates the plate-current pulses that would result from the application of this sine wave to grid No. 1 of the stage shown in the simplified circuit, provided that the signal had the same phase as sine wave A. If the signal shown in Fig. 7-1B had the same phase as that of sine wave B, there would be no change in plate-current pulses, because the sine wave would always pass through zero during the time the tube conducts.

In order to recover the amplitude variations of the signal shown by waveform B, the chrominance signal must be fed into another synchronous demodulator. The suppressor grid in this stage would be pulsed during time 3 instead of time 2. At time 3, sine wave B reaches its peak and sine wave A is at zero. The plate-current pulses would then be affected by the modulation of sine wave B, and the amplitude variations of the signal would be detected.

DEMODULATION USING A CW SIGNAL

Demodulation of the chrominance signal is not as simple as the preceding discussion implies. It would not be practical to use a manual switching arrangement for the grid No. 3 voltage. In actual practice, grid No. 3 is controlled automatically by a cw reference signal which is developed in the color-sync section of the receiver. The demodulator stage is biased in such a way that the tube conducts only when the reference signal reaches the peak of its positive cycle.

A synchronous-demodulator circuit employed in color receivers is shown in Fig. 7-2. The control grid for each stage receives the same chrominance signal from the output of the bandpass amplifier.

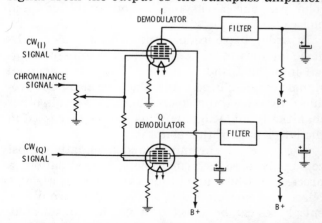

Fig. 7-2. A synchronous-demodulator circuit employed in a tube-type color receiver.

Grid No. 3 of each demodulator is supplied with a cw reference signal from the output circuit of the color-sync section. The cw reference signal that is present on grid No. 3 of each stage is at a constant amplitude and frequency; however, the phase difference between the two reference signals is 90 degrees. The chrominance signal at the input of each stage is constantly varying in phase and amplitude with changes of hue and saturation of the colors in the transmitted scene.

The reference signal applied to the I demodulator may be in phase or 180 degrees out of phase with the in-phase portion of the chrominance signal, and the reference signal applied to the Q demodulator may be in phase or 180 degrees out of phase with the quadrature portion of the chrominance signal. Proper demodulation will result in either case, but the polarity of the output signal will be reversed if the phase of the cw reference signal is changed 180 degrees.

Let us examine the manner in which the demodulators detect the amplitude variations of the two portions of the chrominance signal. The waveforms for the cw reference signals that appear on the No. 3 grids of the demodulators are shown at the top of Fig. 7-3. As shown in the drawing, the conduction time of each demodulator occurs during the positive peak of its respective reference signal. The sampling period of the demodulator is very short. At any other time, the stages are not conducting. Since current is allowed through either tube only during its respective sampling period, detection can occur only during such time.

The chrominance signal that appears at the control grids of the demodulators can assume the shape of any one of the waveforms shown by the solid lines (waveforms A, B, and C) below the reference signals in Fig. 7-3. The waveforms shown by the dashed lines represent the two modulated signals that have been combined at the transmitter to form the chrominance signal. Since the chrominance signal is always a sine wave, its amplitude will increase and decrease as the sine of an angle will increase and decrease. At 0°, the amplitude of a sine wave is zero; at 30 degrees, the amplitude is 50 percent of the peak value; at 60 degrees, the amplitude is 86.6 percent of the peak value; and at 90 degrees, the amplitude is 100 percent of the peak value.

Waveform A in Fig. 7-3 shows a chrominance signal that lags the I reference signal by 45 degrees and leads the Q reference signal by 45 degrees. It can be seen that the amplitude of wave-

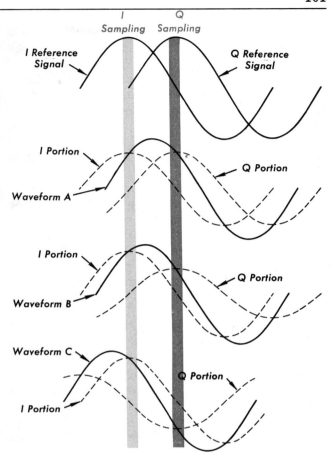

Fig. 7-3. Waveforms showing that the amplitude of the chrominance signal at the time of sampling is the same as the peak amplitude of that portion which is detected.

form A during the sampling period of either demodulator is about 70 percent of its peak amplitude. Note that the points at which this chrominance signal is sampled are the same ones at which its two portions reach their peak amplitudes; therefore, during the sampling period of the I demodulator, the voltage on the control grid of this tube has an amplitude equal to the peak value of the in-phase portion of the chrominance signal. During the sampling period of the Q demodulator, the voltage on the control grid of this tube has an amplitude equal to the peak value of the quadrature portion of the chrominance signal.

If the chrominance signal takes the form of waveform B, its phase will lag that of the I reference signal by 30 degrees and lead that of the Q reference signal by 60 degrees. During the sampling period of the I demodulator, the value of this signal is approximately 87 percent of the value of its positive peak. As shown in Fig. 7-3, the value of the voltage on the control grid of each demodulator during the respective sampling period of the tube corresponds to the peak amplitude of that portion of the chrominance signal which the stage is designed to detect.

It is also possible for the chrominance signal to be passing through the negative portion of its cycle during the sampling period of one or both of the demodulators. For example, the sine wave shown by waveform C in Fig. 7-3 has a value equal to 87 percent of the positive peak when this signal is sampled by the I demodulator; however, the value of this signal during the sampling period of the Q demodulator is equal to 50 percent of the negative peak. Note that the quadrature portion of the signal shown by waveform C has reached its negative peak at this time. Since the presence of a negative voltage on the control grid of the Q demodulator increases the tube bias, the average current through the tube is less than when no signal is present on the control grid.

An overall analysis of Fig. 7-3 should make it clear that the current through the I demodulator is proportional to the peak amplitude of the in-phase portion of the chrominance signal and is not affected by the amplitude variations of the quadrature portion. The current through the Q demodulator is proportional to the peak value of the quadrature portion and is not affected by the amplitude variations of the in-phase portion of the chrominance signal.

As long as no signal voltage is present on the control grid of a demodulator tube, the current during the sampling period will have a nominal value. When a chrominance signal is present on the control grid, the current during the sampling period of a demodulator will increase if that portion of the chrominance signal recovered by the stage has the same polarity as the reference signal applied to the tube. Tube current will decrease if the polarity of this signal is opposite to that of the reference signal. If the signal applied to a demodulator stage does not contain that portion of the chrominance signal which the demodulator is designed to recover, tube current will be the same as when no signal is present on the control grid.

PHASE AND AMPLITUDE OF THE CHROMINANCE SIGNAL

Before proceeding further, let us consider the color-phase diagram shown in Fig. 7-4. Vectorially, it shows the phase relationship between the color burst and the chrominance signal when the camera scans a primary or complementary color. The amplitudes and polarities of the in-phase and

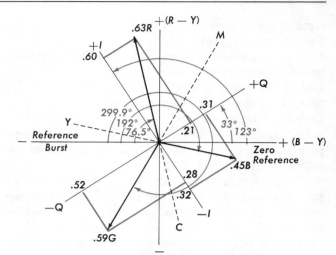

Fig. 7-4. Color-phase diargam.

quadrature signals needed to produce a chrominance signal having any of these phase angles are also shown.

The phase diagram in Fig. 7-4 further shows that when a fully saturated red is scanned by the color camera, a chrominance signal that lags the color burst by 76.5 degrees and that has a peak amplitude of .63 of unity is produced. It can be seen that such a chrominance signal is produced when the amplitude of the in-phase portion is .6 and the amplitude of the quadrature portion is .21.

Scanning of a fully saturated cyan causes the formation of a chrominance signal that lags the color burst by 256.5 degrees and that has a peak amplitude of .63. The chrominance signals formed when red or cyan is being scanned will therefore have the same amplitudes but opposite polarities. This is logical because cyan is the complement of red. As shown by the phase diagram, the vector that designates the amplitude and phase of the chrominance signal produced while scanning the color cyan is opposite to the vector that designates the phase and amplitude of the chrominance signal produced while scanning the color red. The chrominance signal for cyan consists of in-phase and quadrature portions that have negative polarities.

Scanning the color green results in the formation of a chrominance signal that lags the color burst by 299.9 degrees and that has a peak amplitude of .59. Scanning the color blue results in the formation of a chrominance signal that lags the color burst by 192 degrees and that has a peak amplitude of .45. Note on the phase diagram that the values of the I and Q vectors for green are −.28 and −.52 and for blue are −.32 and +.31, respectively.

SAMPLING OF THE CHROMINANCE SIGNAL

The sampling process of the I and Q demodulators for each of the primary and complementary colors will now be investigated. Chrominance signals that are produced as a result of scanning each of these colors are shown in Fig. 7-5. The waveforms for the cw reference signals are again shown at the top of the illustration. In order to reproduce the primary colors, there must be an output signal from both of the demodulators for each color. This means that portions of the original color signals from the output of the color camera are used to form the I signal and that the remaining portions are used to form the Q signal. These two signals modulate quadrature portions of the chrominance subcarrier to form the chrominance signal; therefore, the chrominance signal at the input of the demodulators in the receiver must be sampled at two points during its cycle in order to detect the amplitude variations of the original modulated signals.

Section A of Fig. 7-5 shows the waveform of the chrominance signal produced when the camera scans the primary color red and the one produced when the camera scans the complementary color cyan. The waveform produced while cyan is being scanned is shown in dashed lines, because it is actually a different signal from that produced while red is being scanned. It is shown in this way to illustrate that the chrominance signal which results while red is being scanned is of the same amplitude as that which results while cyan is being scanned, but it is of opposite polarity. In the case of red, the chrominance signal lags the I reference signal by 19.5 degrees and leads the Q reference signal by 70.5 degrees. It is sampled by the I demodulator at an amplitude of .6 and by the Q demodulator at an amplitude of .21. Each time a chrominance signal having a phase relationship like that shown in Fig. 7-5 appears, the color red is being represented electrically. When the chrominance signal is of the same amplitude but of the opposite phase, the color cyan is being represented electrically.

Section B of Fig. 7-5 shows the phase and amplitude of the chrominance signal for the primary color green and for the complementary color magenta. For green, the chrominance signal leads the I reference signal by 117 degrees and lags the Q reference signal by 153 dgrees. The chrominance signal is sampled at an amplitude of −.28 by the I demodulator and at an amplitude of −.52 by the Q demodulator. For the color magenta, the chrominance signal is sampled at the same two amplitudes but at the opposite polarity of the primary color green.

The waveforms in section C of Fig. 7-5 are the chrominance signals for the primary color blue and the complementary color yellow. The signal for blue lags the I reference signal by 135 degrees and lags the Q reference signal by 45 degrees. It is sampled at an amplitude of −.32 by the I demodulator and at +.31 by the Q demodulator. As was true in the foregoing cases, the chrominance signal for yellow has the same amplitude as that for blue but has the opposite polarity; consequently, for yellow, the chrominance signal is positive when sampled by the I demodulator and negative when sampled by the Q demodulator.

DEMODULATION WITH ONLY I OR ONLY Q PRESENT

Shown in Fig. 7-6 are waveforms of chrominance signals that will cause a signal to be produced at the output of one demodulator and none at the output of the other. In section A of Fig. 7-6, the waveforms of the two cw signals are shown so

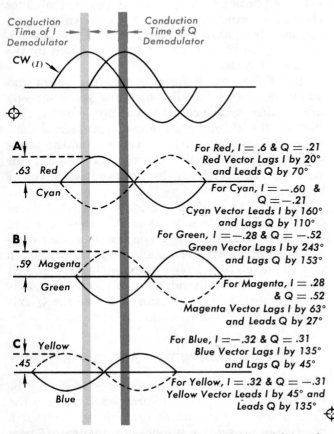

Fig. 7-5. Sampling process of the I and Q demodulators for each of the primary colors and their complements.

that a comparison of the sampling times can be made. Waveforms designated as B, C, D, and E are the sine waves of different chrominance signals on the input grids of the demodulators. Sine wave B is in phase with the I reference signal; therefore, a signal is produced at the output of the I demodulator. There will be no output signal from the Q demodulator because the chrominance signal is going through zero during the sampling time of the Q demodulator.

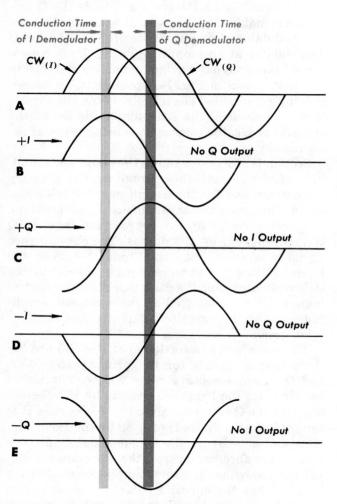

Fig. 7-6. Sampling process of the chrominance signal containing only the I and Q signals.

Sine wave D is a chrominance signal that is 180 degrees out of phase with the I reference signal; therefore, the polarity of the output signal of the I demodulator will be opposite that of the output signal when sine wave B is being demodulated. There will be no output signal at the plate of the Q demodulator because sine wave D goes through zero during the sampling time of the Q demodulator.

Sine wave C is a chrominance signal that is in phase with the Q reference signal and is 90 degrees out of phase with the I reference signal; therefore, it reaches its peak during the sampling time of the Q demodulator. There will be an output signal produced by the Q demodulator, but none will be produced by the I demodulator. The same thing is true in the case of sine wave E, except that the phase of this signal is opposite that of the Q reference signal. Sine wave E is at its negative peak during the sampling time of the Q demodulator, and the polarity of the output signal of the Q demodulator will be the reverse of that produced during the demodulation of sine wave C. As in the case of sine wave C, sine wave E is passing through zero during the sampling time of the I demodulator; therefore, no output signal is produced by the I demodulator at this time.

The waveforms shown in Fig. 7-6 are representative of chrominance signals in which only I or Q demodulation is present. This condition exists when certain colors are being transmitted. Only an I signal is formed when the colors that lie on the I axis (orange and a near cyan) are scanned by the camera. With the vectors plotted as shown in Fig. 7-4, the phase and amplitude of the chrominance signal for orange are shown by the positive I vector, and the phase and amplitude of the chrominance signal for the near cyan are shown by the negative I vector.

Only a Q signal is formed when the colors that lie on the Q axis (reddish-blue and yellow-green) are scanned by the camera. The phase and amplitude of the chrominance signal for reddish-blue are shown by the positive Q vector, and the phase and amplitude of the chrominance signal for yellow-green are shown by the negative Q vector in Fig. 7-4.

For the sake of simplicity, the colors that lie on the I axis are commonly referred to as orange and cyan, and the colors that lie on the Q axis are commonly referred to as green and magenta. Although this nomenclature is not completely accurate, it has been universally adopted. In fact, the I axis can be called the orange-cyan axis, and the Q axis can be called the green-magenta axis.

DEMODULATION WITH I AND Q PRESENT

The in-phase and quadrature portions may be combined in any one of several ways to form the chrominance signal, which can consist of either a positive or a negative in-phase or quadrature portion. It can also consist of combinations of the two

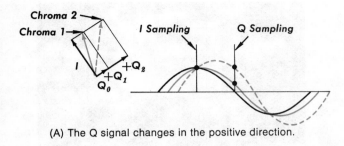

(A) The Q signal changes in the positive direction.

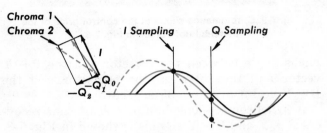

(B) The Q signal changes in the negative direction.

Fig. 7-7. Sampling process when the I signal remains constant and when the Q signal varies in positive and negative directions.

portions—both positive, both negative, or one positive and one negative in polarity.

To further illustrate the fact that the chrominance signal can be made up of any combination of the in-phase and quadrature portions, consider a case in which the amplitude of the in-phase portion remains constant and the quadrature portion varies in amplitude. Shown in Fig. 7-7 are the resultant vector relationships when the amplitude of the in-phase portion remains constant and the amplitude of the quadrature portion increases. The illustration shows the resultant vectors when the quadrature portion increases in amplitude from zero to the values shown by the vectors Q_1 and Q_2.

The I vector shows the phase and amplitude of the chrominance signal for orange. The resultant vector, called chroma 1, shows the phase and amplitude of the chrominance signal for red, and chroma 2 shows the phase and amplitude of the chrominance signal for bluish-red. These three chrominance signals are shown by the waveforms alongside the drawings of the vectors. When the chrominance signal comprises the in-phase portion only, its waveform is like that shown by the dark solid line. This chrominance signal would be in phase with the reference signal and would be sampled at the peak of the wave.

The waveform of the chrominance signal that consists of the I and Q_1 portions is shown by a light solid line. This chrominance signal lags and is greater in amplitude than the sine wave shown

by the waveform in the dark solid line. Note that this signal is sampled by the I demodulator at the same amplitude as the signal shown by the waveform in the dark solid line, and has a positive value when sampled by the Q demodulator.

The waveform of the chrominance signal that consists of the I and Q_2 portions is shown by the dashed line. The chrominance signal is still sampled by the I demodulator at the same amplitude as the chrominance signals shown by the other two waveforms, but has a higher amplitude when sampled by the Q demodulator.

Fig. 7-7B shows the waveform of the chrominance signal when the quadrature portion has a negative polarity and increases in amplitude while the amplitude of the in-phase portion remains constant. The resultant vector (chroma 1) can represent the phase and amplitude of a chrominance signal for yellow-orange, and chroma 2 can represent them for yellow. The waveform of the chrominance signal when the in-phase portion is maximum and when the quadrature portion is zero is the solid line of Fig. 7-7B. The waveform of a chrominance signal that consists of I and negative Q_1 portions is shown by the light solid line in Fig. 7-7B. This signal is sampled by the I demodulator at the same amplitude as the signal shown by the dark solid line, and it has a negative value when sampled by the Q demodulator. The signal shown by the light solid line leads in phase and has a higher amplitude than the signal shown by the dark solid line.

The waveform of a chrominance signal, which consists of I and negative Q_2 portions, is shown by the dashed line. The signal shown by this waveform leads those shown by the other two waveforms and is higher in amplitude. When sampled by the Q demodulator, this signal has a greater negative value than the chrominance signal shown by the light solid line. All three chrominance signals are sampled at the same amplitude by the I demodulator.

An important point that must be remembered about the chrominance signal is that it is a sine wave, and by varying the amplitude and phase of the signal, any combination of values (whether plus or minus) of the I and Q signals can be represented electrically. The variety of chrominance signals shown by the waveforms in Fig. 7-7 should illustrate this fact.

While the demodulator circuits in most early color receivers were designed to recover the I and Q signals, modern color receivers usually demodulate on some other axes such as R − Y and B − Y.

In the remainder of this chapter, we will discuss the more-common demodulator circuits now found in tube-type and solid-state color receivers.

R − Y AND B − Y DEMODULATOR CIRCUITS

Some receivers are designed to recover R − Y and B − Y signals instead of I and Q signals. This is possible, since the I and Q signals are made up of specific proportions of the color-difference signals R − Y and B − Y. The following equations (which first appeared in Chapter 3) for the I and Q signals point out this fact.

$$E_I = .74(E_R - E_Y) - .27(E_B - E_Y)$$
$$E_Q = .48(E_R - E_Y) + .41(E_B - E_Y)$$

If the cw reference signals developed by the color-sync section have specific phase relationships with the color burst, the R − Y and B − Y signals will be recovered from the chrominance signal by the demodulators. Note from the color-phase diagram of Fig. 7-4 that the phase of the reference signal used to recover the B − Y signal is in phase with the zero-reference axis and that the phase of the reference signal used to recover the R − Y signal is displaced 90 degrees. The

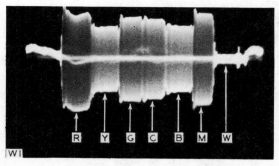

Fig. 7-9. Chrominance signal at the control grids of the demodulator stages in Fig. 7-8.

phase angle between the I vector and the R − Y vector and that between the Q vector and the B − Y vector are each 33 degrees.

A synchronous-demodulator circuit that recovers R − Y and B − Y signals is shown in Fig. 7-8. This circuit is used in a tube-type color receiver. Two 6GY6 pentode tubes are employed as the demodulators. The chrominance signal is applied to the control grids, and the cw reference signals are applied to the suppressor grids. Fig. 7-9 shows the waveform of the chrominance signal, and Fig. 7-10 shows the waveform of a cw reference signal. (Only one waveform is shown for a cw signal because the two have the same appearance on the oscilloscope.)

The operation of the circuit shown by Fig. 7-9 is similar to the operation of the basic synchronous-demodulator circuit previously discussed. The main difference between the two is that the cw reference signals have different phase angles. This results in output signals that are different from those of the I and Q demodulators because the chrominance signal is sampled at different points. The drawing in Fig. 7-11 shows the sampling points of a chrominance signal during I and Q demodulation and during R − Y and B − Y demodulation. The sine wave is sampled at times 1 and 3 by the I and Q demodulators and at times 2 and 4 by the R − Y and B − Y demodulators.

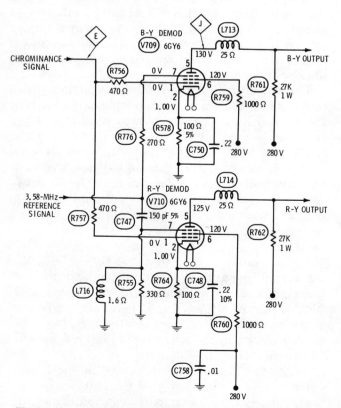

Fig. 7-8. B − Y and R − Y synchronous demodulator circuit used in a tube-type color receiver.

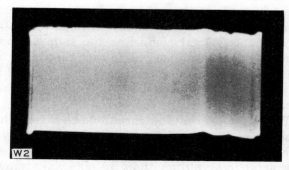

Fig. 7-10. The cw reference signal in the circuit of Fig. 7-8.

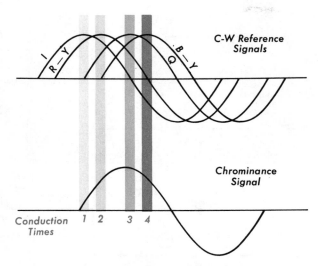

C-W Reference Signals

Chrominance Signal

Conduction Times   1   2   3   4

Fig. 7-11. Comparison of sampling times for the I, Q, R − Y, and B − Y signals.

The 3.58-MHz cw reference signal is applied to the control grid of B − Y demodulator V709 in phase. The cw reference signal applied to the control grid of the R − Y demodulator V710 is shifted 90 degrees by the phase-shift network consisting of C747 and L716.

The signal at the output of V709 is a B − Y signal, and the output of V710 is an R − Y signal. The bandwidths of these two output signals are limited to 0.5 MHz. The bandpass for the B − Y channel is limited by L713, and the bandpass for the R − Y channel is limited by L714. Since both channels have the same bandwidth, the response curve for either channel would appear like that shown in Fig. 7-12.

The limited bandwidth of the demodulator output circuits removes the 3.58-MHz pulses from the output signal. The B − Y output is shown in Fig. 7-13, and the R − Y output is shown in Fig. 7-14. The waveforms shown here are from an NTSC color-bar generator, since the NTSC color-bar pattern was used in the discussion of the makeup of the color-picture signal in Chapter 3. However, the

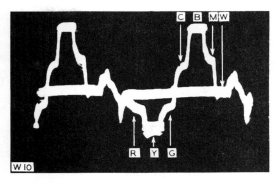

Fig. 7-13. Output signal of the B − Y demodulator in Fig. 7-8.

waveforms usually encountered in service literature are from a keyed-rainbow color-bar generator.

Note that the proper bias for each demodulator stage is established by the 100-ohm resistors in the cathode circuits. Both of the resistors are bypassed by a .22-μF capacitor. The output signal from each demodulator stage in Fig. 7-8 is positive in polarity. A signal with positive polarity is needed in this receiver because the color-difference signals from the demodulators are fed to the control grid of the picture tube.

It has been stated before that the output signals produced by the R − Y and B − Y demodulators are different from the output signals of the I and Q demodulators. For a comparison of the differences, refer to Fig. 7-15 which shows waveforms of the demodulated signals that will be produced when an NTSC color bar such as the one indicated at the top of the illustration is scanned.

Whenever R − Y and B − Y demodulators are used, the output signals will take the form of the top two waveforms in Fig. 7-15. If I and Q demodulators are used, the output signals will take the form of the lower two waveforms in Fig. 7-15. All of these signals are positive in polarity. If the output signals of the demodulators were negative, the waveforms would be opposite in polarity. The

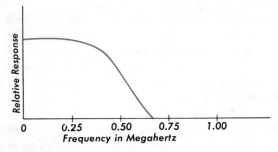

Fig. 7-12. Passband of either the R − Y or the B − Y demodulator circuit in Fig. 7-8.

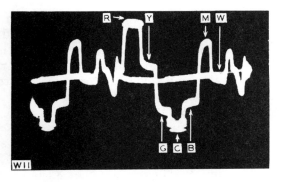

Fig. 7-14. Output signal of the R − Y demodulator in Fig. 7-8.

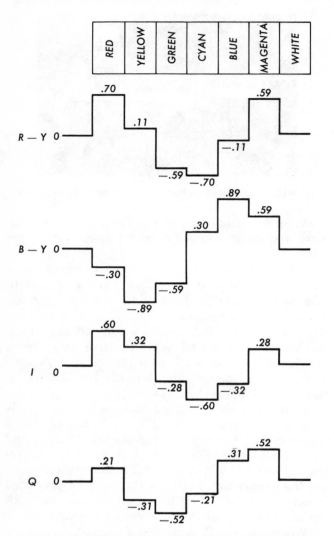

Fig. 7-15. Comparison of R − Y, B − Y, I, and Q signals.

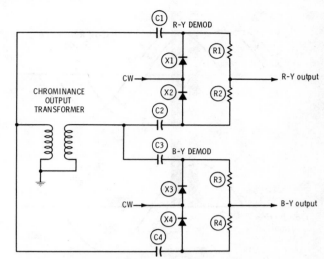

Fig. 7-16. Basic diode-demodulator circuit.

polarity of the signal at the output of a demodulator can be changed from positive to negative by utilizing a cw reference signal of the opposite polarity.

DIODE DEMODULATORS

A basic circuit employing diode demodulators is shown in Fig. 7-16. One pair of diodes is used as the R − Y demodulator, and the other pair is used as the B − Y demodulator. Each demodulator is connected so that one diode will produce an output signal of positive polarity at its cathode and the other diode will produce an output signal of negative polarity at its anode.

In the R − Y demodulator, the chrominance signal is applied to the cathode of X1 and to the anode of X2. The chrominance signal at X1 is 180 degrees out of phase with the chrominance signal at

diode X2. This phase reversal is due to the action of the transformer at the takeoff point of the chrominance signal. The R − Y reference signal from the color-sync section is applied to the anode of X1 and to the cathode of X2. The output signal of the R − Y demodulator is present at the junction of resistors R1 and R2.

The B − Y demodulator is connected in a manner similar to that of the R − Y demodulator. The chrominance signal is applied to the cathode of X3 and to the anode of X4. The signal present at one diode is 180 degrees out of phase with the signal present at the other diode. The B − Y reference signal is applied to the anode of X3 and to the cathode of X4. The output signal of the B − Y demodulator appears at the junction of resistors R3 and R4.

The amplitude of the output signal from each demodulator is dependent upon the amount of current through each diode of the demodulator. Whenever the currents through the diodes are equal, the output will be zero. For example, when both diodes of the R − Y demodulator are conducting equally, a positive voltage will appear at the junction of R1 and the cathode of diode X1 because of the positive charge on C1. A voltage of the same magnitude but of opposite polarity will appear at the junction of R2 and the anode of X2 because of the negative charge on C2. Since the charges on C1 and C2 are equal but of opposite polarity, and since both C1 and C2 are returned to ground through the chrominance output transformer, a zero voltage is present at the junction of the equal-value resistors, R1 and R2. This condition would exist when there is no chrominance signal at the input. Under the same conditions, the B − Y demodulator would perform in the same manner as just described for the R — Y demodulator.

Let us now investigate the action of the diode demodulators during the time that a chrominance signal is being received. For example, let us assume that a chrominance signal which is lagging the R — Y reference signal by 45 degrees is present at the cathode of diode X1 of the R — Y demodulator. The signals that appear at both diodes of the demodulators are shown by the waveforms in Fig. 7-17. The first pair of waveforms shows the signals at diode X1 of the R — Y demodulator. The waveform marked A is the reference signal at the anode, and the one marked K is the chrominance signal at the cathode. The conduction time is the period during which the anode is more positive than the cathode, and is represented by the shaded portion between the sine waves.

The second pair of waveforms in Fig. 7-17 shows the signals at diode X2 of the R — Y demodulator. The reference signal appears at the cathode of this section and is shown by the waveform marked K. It is the same signal that was present at the anode of X1. The waveform marked A is the chrominance signal that appears at the anode of X2. This chrominance signal is 180 de-

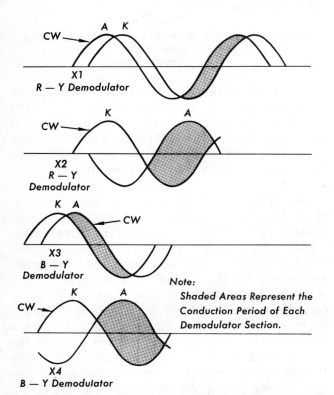

Fig. 7-17. Conduction period of each demodulator when the chrominance signal lags the R — Y reference signal and leads the B — Y reference signal by 45°.

grees out of phase with that labeled K in the first pair of waveforms. The conduction time of section B is the shaded portion between the two sine waves.

It can be seen that diode X2 conducts more heavily than diode X1. This causes the charge on C2 to be greater than that on C1; and because the charge on C2 is negative, a negative voltage is present at the junction of R1 and R2.

The third pair of sine waves in Fig. 7-17 shows the signals present at diode X3 of the B — Y demodulator. The chrominance signal is waveform K, and it appears at the cathode of X3. This chrominance signal has the same phase as that at the anode of X2 in the R — Y demodulator. Waveform A of the third pair is the B — Y reference signal that appears at the anode of X3. This reference signal is advanced 90 degrees in phase with respect to the phase of the R — Y reference signal. The conduction time of diode X3 of the B — Y demodulator is represented by the shaded portion of the third pair of waveforms in Fig. 7-17.

The fourth pair of waveforms in Fig. 7-17 represents the signals that are present at diode X4 of the B — Y demodulator. In this diode, the chrominance signal appears at the anode and the reference signal appears at the cathode. Waveform A is the chrominance signal and has the same phase as the signal that appears at the cathode of diode X1 in the R — Y demodulator. Waveform K is the reference signal. The conduction time of diode X4 of the B — Y demodulator is represented by the shaded portion between the two waveforms.

Diode X4 of the B — Y demodulator conducts more heavily than diode X3. As a result, there is a greater charge on C4 than on C3, and the output signal of the B — Y demodulator will be negative in value.

For the case just assumed, the chrominance signal lagged the R — Y reference signal by 45 degrees, and it led the B — Y reference signal by 45 degrees. The output signal of each demodulator was found to be negative in value.

Let us assume that the chrominance signal leads the R — Y reference signal by 45 degrees and leads the B — Y reference signal by 135 degrees. The conduction period of each diode of the R — Y and B — Y demodulators is represented by the shaded portions between the waveforms in Fig. 7-18. By inspection of the first two pairs of waveforms, it can be seen that diode X2 of the R — Y demodulator conducts more heavily than diode X1. This will produce a negative voltage at the output of the R — Y demodulator.

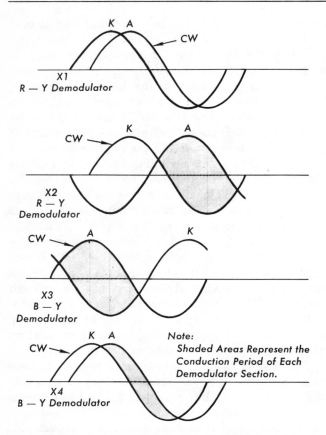

Fig. 7-18. Conduction period of each demodulator when the chrominance signal leads the R — Y reference signal by 45°, and leads the B — Y reference signal by 135°.

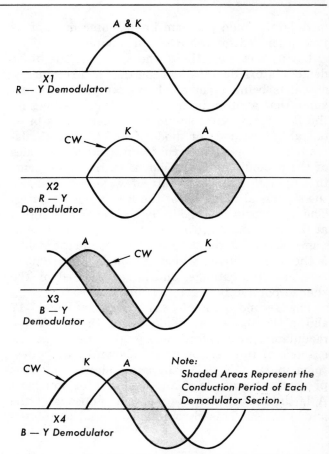

Fig. 7-19. Conduction period for each demodulator when the chrominance signal is in phase with the R — Y reference signal.

The B — Y demodulator will produce an output voltage that is positive, as can be seen from inspection of the last two pairs of waveforms in Fig. 7-18. Diode X3 of the B — Y demodulator conducts more heavily than diode X4; therefore, the charge on C3 will be greater than that on C4, and the resultant output voltage from the B — Y demodulator will be positive.

Let us investigate what happens when the chrominance signal is in phase with the R — Y reference signal. Under this condition, there must be a maximum negative voltage at the output of the R — Y demodulator and zero voltage at the output of the B — Y demodulator. The first two pairs of waveforms in Fig. 7-19 show that diode X1 of the R — Y demodulation will not conduct, since the chrominance signal is in phase with the reference signal. The conduction of diode X2 of the R — Y demodulator will be maximum (as shown by the second pair of waveforms) because the chrominance signal has been changed in phase by 180 degrees. A maximum negative voltage will appear at the junction of R1 and R2, the R — Y output, under this condition.

The third and fourth pairs of waveforms in Fig. 7-19 show that the conduction of diode X3 of the B — Y demodulator is equal to the conduction of diode X4. The charges on C3 and C4 are equal, and the output voltage of the B — Y demodulator will be zero, as if there were no chrominance signal present.

The vector relationship of the signals previously discussed is shown in the diagram of Fig. 7-20. The phase and amplitude of the chrominance signal, which lagged the R — Y reference signal and led the B — Y reference signal by 45 degrees, are shown by the vector marked chroma 1. A chrominance vector at this position on the phase diagram would represent a color near magenta. A chrominance signal having the phase and amplitude represented by this vector is formed when the R — Y and B — Y signals have positive polarities. It may be remembered from the color-phase diagram shown in Fig. 7-4 that the R — Y and B — Y signals had positive polarities when a chrominance signal having this phase angle was demodulated.

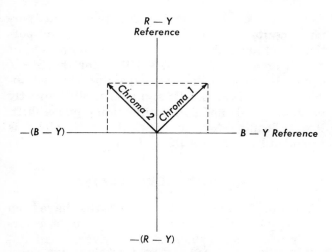

Fig. 7-20. Vector relationship between the chrominance signal and the reference signal.

The vector marked chroma 2 shows the phase and amplitude of the chrominance signal that led the R — Y reference signal by 45 degrees and led the B — Y reference by 135 degrees. This vector represents a yellow-orange. A chrominance signal having the phase and amplitude represented by this vector is formed when the R — Y signal has a

positive polarity and the B — Y signal has a negative polarity.

The R — Y and B — Y demodulators can be made to produce an output signal of a positive or a negative polarity, depending on which is desired. Whenever an output signal of a positive polarity is wanted, the demodulator is keyed with a cw signal having a specific phase angle. When one of a negative polarity is desired, the demodulator is keyed with a cw signal having the opposite phase angle.

An R — Y and B — Y demodulator circuit using diodes is shown in Fig. 7-21. This circuit is employed in a hybrid color receiver. Diodes CR3 and CR4 are used in the B — Y demodulator stage, while diodes CR1 and CR2 are used for the R — Y demodulator stage. The operation of this circuit is the same as that described for the basic circuit shown in Fig. 7-16. However, the B — Y and R — Y outputs from this circuit are negative. That is, they produce signals that are reversed in polarity from the signals used to modulate the chrominance subcarrier at the transmitter. These negative signals are inverted by the color-difference amplifiers (not shown) and they are applied to the con-

Fig. 7-21. Diode B — Y and R — Y demodulator circuit used in a hybrid color receiver.

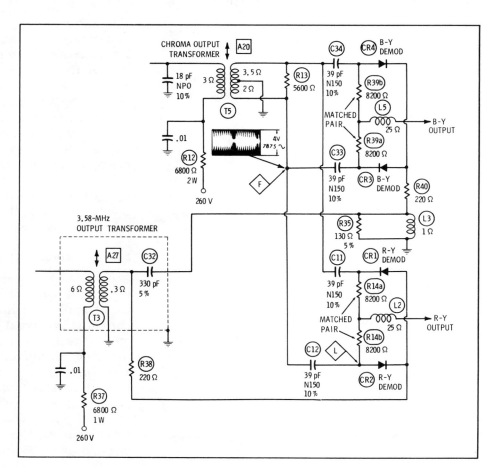

trol grids of the color picture tube with positive polarity.

B − Y, R − Y, G − Y DEMODULATION USING BALANCED DUAL-DIODE DETECTORS

Three color demodulator circuits, one for each color-difference signal, are employed in the RCA circuit shown in Fig. 7-22. The demodulator circuits discussed previously produced only two output signals. This is possible because the third color-difference signal can be obtained by the proper mixture of the other two color-difference signals. For example, the G − Y signal is made up of −.51 (R − Y) and −.19 (B − Y). The use of a separate G − Y demodulator tends to increase the bandwidth of the signal.

The chroma demodulator circuits in Fig. 7-22 are balanced dual-diode detectors. The output of each dual-diode circuit is proportional to both the phase and amplitude of the applied signal.

Two signals are applied to each demodulator circuit—a composite color signal from the bandpass amplifier and the reference signal from the 3.58-MHz oscillator circuit. The phase of the reference signal is shifted a specific amount with respect to the burst signal for each demodulator, extracting the appropriate color-difference signal from the input chroma signal. Circuit action is as follows:

The phase of the 3.58-MHz reference signal applied to the R − Y demodulator is shifted by ca-

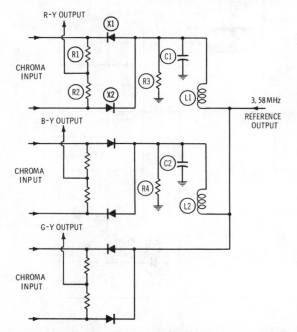

Fig. 7-22. B − Y, R − Y, G − Y color demodulator circuits used in an RCA solid-state chassis.

pacitor C1 and inductor L1. This phase shift permits the R − Y demodulator output to be proportional to the amplitude of the R − Y component of the chroma signal. Phase shifting for the B − Y signal is accomplished by capacitor C2 and inductor L2. The 3.58-MHz signal is applied directly to the G − Y demodulator without any phase shift. Proper loading for the 3.58-MHz cw amplifier is provided by resistors R3 and R4.

HIGH-LEVEL DEMODULATION

Most of the demodulator circuits that have been discussed so far provide, at their outputs, color-difference signals which have relatively small amplitudes. As a result, each of these color-difference signals must be amplified before being applied to the picture tube. Before demodulation, only the chroma signal needs to be amplified, but after demodulation, two to five signals must be amplified. In receivers employing what are termed *high-level* demodulators, the chrominance signal is amplified to the desired level before it is applied to the demodulators. A distinct advantage is thus gained because the need for several amplifier stages and the adjustments associated with these stages are eliminated.

Because of the frequencies involved, it is easier from a design standpoint to provide for amplification of the chrominance signal before demodulation than it is to provide for amplification of the color signals after demodulation. The frequencies of the chrominance signal fall within the range of 2 to 4.2 megahertz, and the frequencies of the color signals after demodulation are in the range of 30 hertz to 1 megahertz.

In the following discussion, the basic operation of high-level demodulators will be explained, and then two circuits that have been used in tube-type color receivers will be covered.

Basic Circuits of High-Level Demodulators

A basic high-level demodulator circuit is shown in Fig. 7-23. A locally generated cw reference signal is applied to the grid of the triode, and the chrominance signal that is to be demodulated is applied to the plate. The reference signal causes the triode to conduct heavily each time the signal goes through a positive peak of voltage. These regular bursts of current endure for only a small portion of each cycle because the triode operates essentially as a Class-C amplifier. Each time the tube conducts, the plate voltage assumes a fixed value of about 25 volts with respect to ground. At

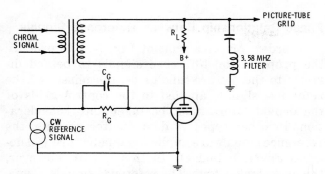

Fig. 7-23. A basic high-level demodulator circuit.

these times, the chrominance signal, which is effectively in series with the plate voltage, will be going through a specific point in its cycle. The location of this point in the cycle will be dependent upon the phase relationship between the cw reference signal and the chrominance signal.

In Fig. 7-24, the pulses of current produced in the tube by the cw reference signal on the grid are shown at the bottom of the illustration. A chrominance signal that is in phase with the cw reference signal will have its positive peaks clamped at 25 volts. A chrominance signal that is 180 degrees out of phase with the cw reference signal will have its negative peaks clamped at 25 volts. The dotted line in the drawing represents the signal waveform that is obtained after the 3.58-MHz portion of the signal is filtered out. This is the waveform of the demodulated color signal.

Only the phase variations of the chrominance signal are shown in Fig. 7-24; the amplitude of the chrominance signal is shown as remaining constant. A change in amplitude would also be detected by the demodulator.

The amplitude of the demodulated output signal is directly proportional to the amplitude of the chrominance input signal; therefore, if a demodulated signal of 90 or 100 volts peak-to-peak is needed, a chrominance input signal of the same

value will produce the desired results. This can be noted from an examination of Fig. 7-24.

Obtaining G − Y Signal From Common Cathode Circuit

The basic circuit that was shown in Fig. 7-23 represents only one triode section of a high-level demodulator stage. At least two triode sections are needed for complete demodulation of the chrominance signal. The basic circuit of a two-triode demodulator is shown in Fig. 7-25. These triodes function as R − Y and B − Y demodulators. It is not necessary to employ three triodes because the third color-difference signal can be obtained by the mixture of the other two color-difference signals. As shown in Fig. 7-25, the G − Y color-difference signal is obtained from the common cathode circuit of the demodulators. This design takes advantage of the fact that a G − Y signal is made up of −.51 (R − Y) and −.19 (B − Y).

The correct operation of the demodulator circuit of Fig. 7-25 depends on three factors: (1) the amplitude of the chrominance signal applied to each plate of the two triodes, (2) the ratios between the plate-load resistors and the common cathode resistor, and (3) the phase relationship between the two reference signals that are applied to the grids of the triodes.

The color-difference signals are attenuated by unequal amounts before transmission in order that the amplitude of the video signal may be kept below that of the sync signal. In order for the correct amplitudes of the color-difference signals to be restored, the amplitudes of the chrominance signals applied to the demodulators must be of

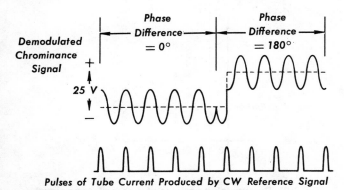

Fig. 7-24. Waveforms illustrating high-level demodulation.

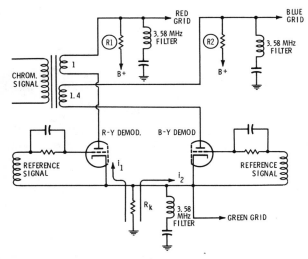

Fig. 7-25. Basic circuit of a two-triode demodulator which employs a common-cathode impedance.

the correct ratio. This is accomplished by using a coupling transformer that has secondary windings with the proper turns ratio. As shown in Fig. 7-25, a turns ratio of 1.4 to 1 is used if the receiver has $B - Y$ and $R - Y$ demodulators. With this specified turns ratio, the chrominance signal applied to the $B - Y$ demodulator will be 1.4 times as great as that applied to the $R - Y$ demodulator.

Selecting Values for Load Resistors

The ratio between the color-difference signals will be correct if the load resistors are of the proper values. The ratio between one plate-load resistor (R1) and common cathode resistor R_k has been found to be 1.96 to 1. The ratio between the other plate-load resistor (R2) and R_k is 5.26 to 1. These ratios have been determined by the following reasoning.

We know that the $G - Y$ signal is obtained from the voltage across the common cathode resistor R_k and that it is made up of $-.51 (R - Y)$ and $-.19 (B - Y)$. We also know the path of the current through both demodulators. As shown in Fig. 7-25, i_1 is the $R - Y$ demodulator current which flows from ground through R_k and through the $R - Y$ triode, and i_2 is the $B - Y$ demodulator current which flows through R_k and through the $B - Y$ triode. From these known facts, the ratio between each of the plate-load resistors, R1 and R2, and the common cathode resistor R_k can be determined as follows:

$$R1 \times i_1 = (R - Y)$$
$$R_k \times i_1 = .51 (R - Y)$$

Then by dividing the first equation by the second, we obtain:

$$\frac{R1}{R_k} = \frac{1}{.51}$$
$$R1 = \frac{R_k}{.51}$$

Then by substituting 1000 ohms for R_k, we obtain:

$$R1 = 1960 \text{ ohms}$$

This is a ratio of 1.96 to 1 between R1 and R_k. The value of R2 can be determined by the same reasoning:

$$R2 \times i_2 = (B - Y)$$
$$R_k \times i_2 = .19 (B - Y)$$
$$\frac{R2}{R_k} = \frac{1}{.19}$$
$$R2 = 5260 \text{ ohms}$$

This is a ratio of 5.26 to 1 between R2 and R_k.

Phase Relationship Between Reference Signals

In order for the circuit of Fig. 7-25 to produce the proper color-difference signals to be fed directly to the picture-tube grids, the phases of the reference signals applied to the control grids of the demodulators must be taken into consideration. In other types of demodulator circuits, the reference signals are applied in quadrature (90 degrees apart in phase). In this circuit, however, this is no longer true. The reference signals have a phase difference of 63.58 degrees. If this phase difference were left at 90 degrees, a $B - Y$ signal produced by the current i_2 flowing through R_k would appear across load resistor R1 and an $R - Y$ signal produced by current i_1 flowing through R_k would appear across R2. Incorrect color-difference signals would therefore be produced and would improperly control the conduction of the guns in the picture tube.

Color-difference signals of the correct amplitudes must be added to the luminance signal in order to reproduce the desired color. For instance, when only red is to be reproduced, the $R - Y$ color-difference signal must have a relative value of .7. This amount of $R - Y$ signal, when added to the luminance signal (which has a value of .3 when only red is being transmitted), will allow full conduction of the red gun. Since the luminance signal is applied equally to all three guns, the $B - Y$ and $G - Y$ signals must each equal a $-.3$ in order to cancel the $+.3$ furnished to each gun by the luminance signal. The green and blue guns should be nonconducting during the reproduction of red.

If the common cathode resistors were not present and if the reference signals were applied in quadrature to the grids of the triodes, the chrominance signal would be demodulated into $R - Y$ and $B - Y$ portions, which are shown by the vector representation in Fig. 7-26. The $R - Y$ portion of

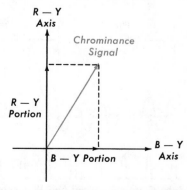

Fig. 7-26. Demodulated portions of a chrominance signal during quadrature demodulation.

the chrominance signal would be developed between the plate side of R1 and the cathode of the R − Y demodulator, and the B − Y portion would be developed between the plate side of R2 and the cathode of the B − Y demodulator.

With the cathode resistor in the circuit, the vector representation of the demodulation action will be alerted. See Fig. 7-27. Because of the small voltage drop E_{Rk} across R_k, the phase of the voltage across the load resistor of each tube will be changed. A small negative vector E_{Rk} is shown added to the R − Y vector to form the resultant vector E_1. This is the new voltage from the plate side of R1 to the cathode of the R − Y demodulator. The small negative vector E_{Rk} is also added to the B − Y vector to form the resultant vector E_2. This is the new voltage from the plate side of R2 to the cathode of the B − Y demodulator. The voltages from the plate sides of the load resistors to the cathodes of the demodulators are no longer R − Y and B − Y but are the two voltages E_1 and E_2. Therefore, for correct demodulation of the chrominance signal, the reference signals applied to the grids must correspond to the phase angles of the vectors E_1 and E_2. When this is done, a pure R − Y signal is formed across plate-load resistor R1, a pure B − Y signal is formed across plate-load resistor R2, and a pure G − Y signal is formed across the cathode resistor R_k.

The vector drawing of Fig. 7-28 shows that the new phase angle of the R − Y reference signal is in quadrature with the vector that represents the blue chrominance signal, and that the new phase

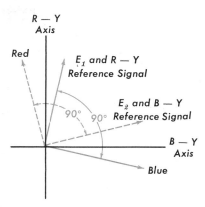

Fig. 7-28. Relationship between the new phase angles of the cw signals and the red and blue vectors.

angle of the B − Y reference signal is in quadrature with the vector that represents the red chrominance signal. With this arrangement and with a chrominance signal representative of the color red being received, an output from only the R − Y demodulator will be produced. There will be no output from the B − Y demodulator. The reverse is true when a chrominance signal representative of the color blue is being received. Demodulation will be performed solely by the B − Y demodulator.

Correct Signals at Guns of Picture Tube

Let us see how the correct signals are applied to the guns of the picture tube under specific conditions. Assume that a red bar is being transmitted and that the incoming signal is such that a signal level of one volt is being produced at the red grid of the picture tube. Under these conditions, we know that the following relative values of signals must be present:

Y = .3
R − Y = .7
B − Y = −.3
G − Y = −.3
B = 0
G = 0
R = 1

From the standard expression for the luminance signal, Y = .3R + .59G + .11B, we know that the luminance signal will equal .3 when only red is being transmitted because the values for blue and green are equal to zero. This .3 value of luminance is applied to all three guns of the picture tube. Therefore, for full conduction of the red gun, an R − Y color-difference signal with a value of .7 must be supplied from the R − Y demodulator. At the same time, a B − Y color-difference signal with

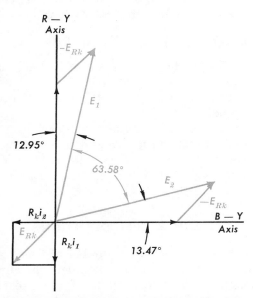

Fig. 7-27. Alteration of the demodulation when the common-cathode impedance is employed.

a value of −.3 must be supplied to the blue gun, and a G − Y color-difference signal with a value of −.3 must be supplied to the green gun so that the luminance signal will be cancelled.

There will be an R − Y color-difference signal of .7 produced across R1 in Fig. 7-25 because (1) the load resistors are in the correct ratio, (2) the chrominance signals driving the demodulators are in the correct ratio, and (3) the reference signals applied to the grids have the correct phase relationship.

The next area of investigation concerns the way in which the circuit of Fig. 7-25 can produce B − Y and G − Y signals with values of −.3. From Fig. 7-28, we can see that the value of voltage E_2 is zero during the transmission of a red bar. This means that there is no difference in signal potential between the plate side of load resistor R2 and the cathode of the B − Y demodulator; therefore, the signal voltage developed across R_k must be equal to the signal voltage developed across R2. In other words, the value of the G − Y signal must equal that of the B − Y signal.

Referring to the relationship

$$G - Y = -.51 (R - Y) -.19 (B - Y),$$

we can show that G − Y is equal to −.3. By substituting G − Y for B − Y and the value of .7 for R − Y in this equation just mentioned, we obtain:

$$G - Y = -.51 (.7) -.19 (G - Y)$$
$$1.19 (G - Y) = -.357$$
$$G - Y = \frac{-.357}{1.19} = -.3$$

B − Y, being equal to G − Y, will also be equal to −.3 during the transmission of a red bar.

If a blue bar were being transmitted, the reverse would be true. Demodulation would be restricted to the B − Y demodulator. The same procedure as the foregoing could be followed to prove that the correct signals are applied to the guns of the picture tube during the transmission of a blue bar.

Obtaining B − Y Signal From Plate Circuits

A variation in the design of a high-level demodulator circuit is the basic circuit shown in Fig. 7-29. Demodulation of the chrominance signal is performed in the same manner as it is by the basic circuit that was shown in Fig. 7-25. The main differences between the two circuits are in the method by which the chrominance signal is shunt fed to the demodulators, and in the method by which the third color-difference signal is obtained.

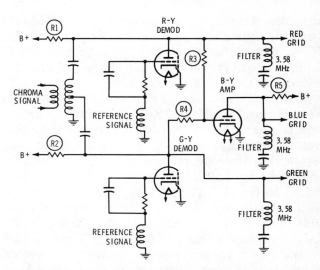

Fig. 7-29. A basic two-triode demodulator circuit that deos not employ a common-cathode impedance.

In the circuit of Fig. 7-25, the G − Y signal is obtained from the common cathode circuit of the demodulator stages. In the circuit of Fig. 7-29, the two triode circuits are designed as R − Y and G − Y demodulators. The B − Y signal is obtained by mixing, in the correct proportion, the R − Y and G − Y signals at the output of the demodulators. Then the signal is passed through a stage of amplification.

The R − Y signal is applied to the grid of the B − Y amplifier through resistor R3. The G − Y signal is applied through R4. The values of these resistors are such that the correct amounts of the R − Y and G − Y signals are mixed together to produce the B − Y signal. The ratio between R3 and R4 is governed by the expression:

$$B - Y = -2.73 (R - Y) -5.36 (G - Y)$$

With R3 and R4 in the ratio of approximately 1 to 2, the R − Y signal and G − Y signal will be mixed (in the correct proportion) to produce the B − Y signal.

The demodulators in the circuit of Fig. 7-29 demodulate along the R − Y and G − Y axes. The relationship of the G − Y axis to the R − Y and B − Y axes is shown by the vector diagram in Fig. 7-30.

X AND Z DEMODULATION

It has been shown that a number of types of demodulators can be used to demodulate the chrominance signal. The phase angle of the axes upon which demodulation takes place must be such that three signals R − Y, B − Y, and G − Y can be

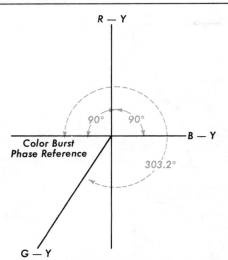

Fig. 7-30. The relative position of the G — Y vector with respect to the R — Y and B — Y vectors.

evolved from the two signals developed by the demodulators.

Fig. 7-31 shows the circuit diagram of the demodulator used in an RCA tube-type chassis. This demodulator operates upon axes that have been arbitrarily called X and Z. The X cw signal and the Z cw signal are obtained from a network in the secondary circuit of transformer L30. The phase angle between the X axis and the Z axis is predetermined by the combination of inductance, capacitance, and resistance, and the angle can change only if these components change value or become defective.

The X cw signal and the Z cw signal drive grids 3 of V23 and V21 respectively, providing the gating signal for the demodulators. A chrominance signal from the chroma-bandpass amplifier is applied to grids 1 of the X and Z demodulators.

A demodulated X signal and a demodulated Z signal appear at the plates of the demodulators. These signals have an increased amplitude due to the amplification of the sampled chroma signal. The amplitude of the chroma signal on the grids of the two demodulators, as shown by the inset waveform, has a peak-to-peak amplitude of 10 volts. The signal at the output of the demodulators has a peak-to-peak amplitude of about 25 volts.

The demodulated signals appearing at the plates of V21 and V23 will contain unwanted 3.58-MHz pulses. These pulses are bypassed to ground by the 33-pF capacitors, C134 and C138, in the plate circuit of each demodulator. The Z signal is then passed through coil L33, and the X signal is passed through coil L32. These coils block the passage of the 3.58-MHz pulses and provide the Z and X signals as shown by the inset waveforms.

The Z signal is applied to the grid of the B — Y amplifier, and the X signal is applied to the grid of the R — Y amplifier. The third signal, G — Y, is then derived from a combination of these two signals.

SOLID-STATE X AND Z DEMODULATION

A solid-state X and Z demodulator circuit is shown in Fig. 7-32. To effectively demodulate the chroma sidebands, the X and Z demodulator circuits provide synchronous detection of these signals with a 3.58-MHz reference injection voltage. The amplitude of the reference signal is several times the magnitude of the chroma-input signal voltage and supplies large-amplitude 3.58-MHz pulses in the emitter circuits of the demodulators. A phase-shift network, comprising inductor L1 and capacitor C1, shifts the phase of the 3.58-MHz signal applied to the Z demodulator approximately 90 degrees. This phase shift provides the most faithful color reproduction.

The phase and amplitude of the chroma signals applied to the bases of the demodulators affect the average amplitude of the pulses present in the collector circuit of each demodulator. These pulses will go somewhat less positive (to approximately one-half the B+ voltage). If the incoming chroma signal is in phase with the 3.58-MHz reference signal, the collector pulses will drop to less than one-half the B+ amplitude. If the incoming chroma signals are out of phase with the reference signal, the collector pulses will not drop to one-half the B+ value. If the incoming chroma signals are 90 degrees out of phase with the reference voltage, part of the collector pulses will fall below the nominal one-half B+ value, and part of the collector pulse will go above the one-half B+ level. This action produces an average of "zero change" in the collector pulses. A low-pass filter network, comprising capacitor C2 and inductor L2 in the X demodulator and capacitor C3 and inductor L3 in the Z demodulator, functions to filter out the 3.58-MHz pulses. As a result of this filtering of the collector signal, only the color information remains. The information is then capacitance-coupled to the color-difference amplifier.

RED, BLUE, AND GREEN DIODE DEMODULATORS

In the circuit shown in Fig. 7-33, three demodulators detect primary red, blue, and green color video signals. A dual-diode phase comparator cir-

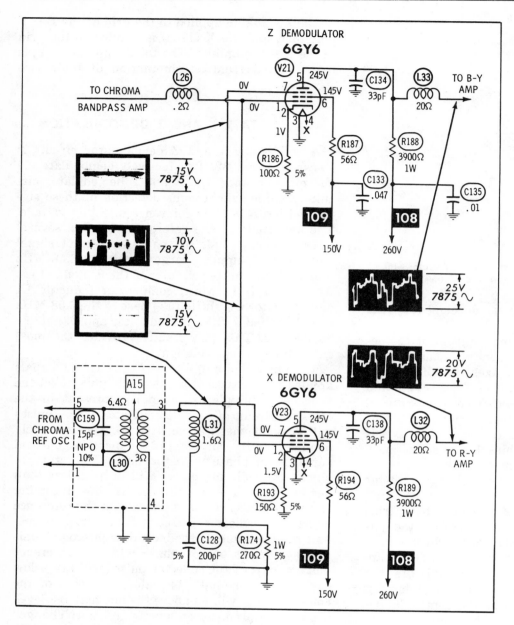

Fig. 7-31. Tube-type X and Z demodulator circuit.

cuit is utilized for each of the three colors. Two discrete signals are compared in phase at each demodulator. One signal is the color information contained in, or derived from, the transmitted signal; the phase of this signal varies with changes in hues. The other signal is the 3.58-MHz reference voltage (reinserted carrier) developed by the reference oscillator. When an in-phase relationship exists between these two signals (Fig. 7-34A) at any given demodulator, a maximum output voltage is developed of such a polarity as to turn on the related crt gun.

A 180-degree out-of-phase relationship between these two signals (Fig. 7-34B) at any given demodulator results in a maximum output voltage of such polarity as to turn off the related crt gun. A 90-degree, or quadrature, relationship between these two signals (Fig. 7-34C) at any given demodulator produces zero voltage output. Instantaneous phase differences between the two signals at any given demodulator that result in output voltages which are not of sufficient magnitude to turn the related crt gun off or on are referred to as intermediate phase angles between the two signals.

The red, blue, and green video signals are demodulated directly by comparing the phase relationship of the reinserted carrier (3.58-MHz reference signal) with the total color signal rather than the individual R − Y, G − Y, and B − Y components of the color signal.

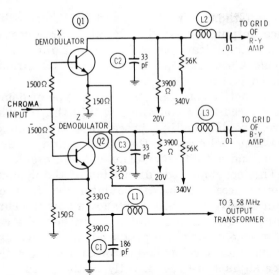

Fig. 7-32. Solid-state X and Z demodulators used in a Sylvania color chassis.

There are a number of advantages to this type of demodulation: color-difference amplifiers are not needed, no dc component is lost, and a uniform demodulation of composite color signals, including

brightness, tends to eliminate matrixing circuit problems.

IC DEMODULATORS

Most late-model color television receivers employ ICs in many of the signal-processing circuits. Fig. 7-35 shows an IC color demodulator circuit used by Zenith. The circuit shown here is actually a portion of a plug-in module that includes the second chroma amplifier. The chroma signal from the chroma level control is fed to pin 2 of the IC through capacitor C116. The green, blue, and red color-difference signals are output at pins 7, 8, and 9 of the IC.

The 3.58-MHz reference signal is applied to pins 4 and 5 of the IC. The proper demodulation angle for this circuit is 104 degrees. This demodulation angle is achieved by shifting the phase of the 3.58-MHz reference signal applied to pin 5 of the IC. Components L702, C712, and R214 provide this shift.

Integrated circuit IC1 contains a pair of synchronous demodulators and a matrix circuit to

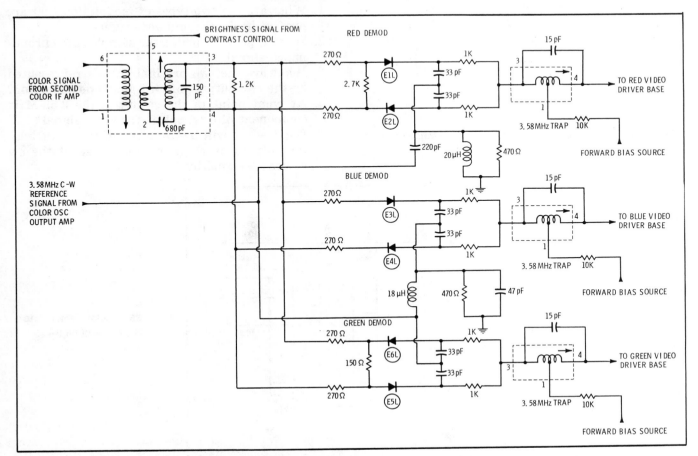

Fig. 7-33. Red, green, and blue demodulator circuit used in a Motorola color receiver.

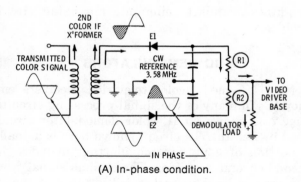

(A) In-phase condition.

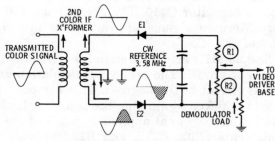

(B) 180° out-of-phase condition.

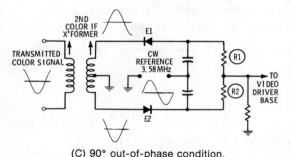

(C) 90° out-of-phase condition.

Fig. 7-34. Phase relationships between the reference oscillator and the transmitted color signal.

provide the G − Y, R − Y, and B − Y color-difference output signals. The IC also furnishes additional amplification for the chroma input signal. A block diagram of the internal IC circuit is shown in Fig. 7-36.

The active components for this type of demodulator circuit are contained within the IC. Therefore, troubleshooting procedures are limited to checking voltages on the IC pins and observing input and output waveforms with an oscilloscope.

This concludes the discussion of the demodulation process and the demodulator circuits used in color receivers. It has been shown how the chrominance signal is demodulated and how the color-difference signals are produced. In receivers that employ high-level demodulators, these color-difference signals are fed directly to the picture tube. When low-level demodulation is employed, the color-difference signals are amplified and appear at the input of the matrix section which will be discussed in the next chapter.

QUESTIONS

1. What are the two types of modulation that are present in the chrominance signal?
2. What type of demodulator samples the chrominance signal?
3. What are the two original signals represented by the two outputs of an I and Q demodulator?
4. At what point in the circuit is the 3.58-MHz component filtered out of the color signal?
5. What effect will be produced at the demodulator outputs by changing the phase of the cw reference signal by 180 degrees?

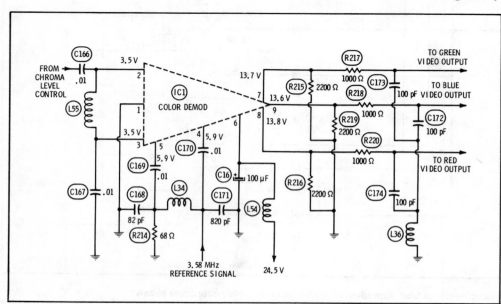

Fig. 7-35. A color-demodulator circuit using an IC.

6. What is the phase relationship between the following:

    (a) The I and Q reference signals?

(b) The R − Y and B − Y reference signals?

(c) Two quadrature signals?

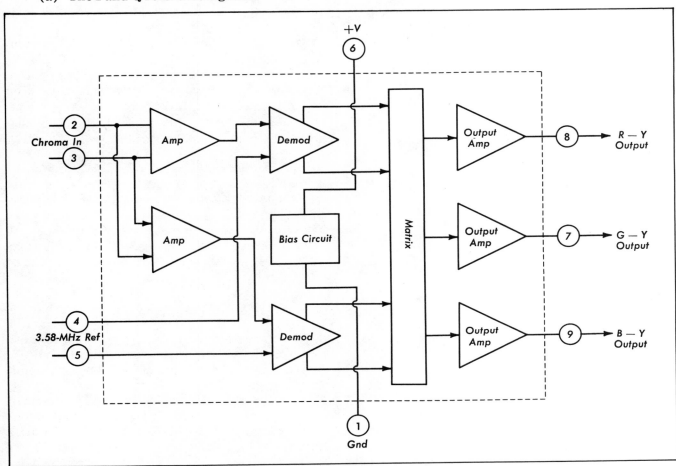

Fig. 7-36. Functional block diagram of the IC in Fig. 7-35.

Chapter 8

The Matrix Section

At this point in the discussion of color receiver circuits, the two types of video signals have been described. These are the luminance signal and the color-difference signal at the output of the chrominance demodulators. It is the function of the matrix section to combine these signals in the correct proportions so that the three color signals are produced. These color signals must correspond to those that appear at the output of the color camera—one for red, one for green, and one for blue. The color signals are amplified and applied to the picture tube where they reproduce the hues of an image in terms of the three primary colors.

In order to understand how the colors of a scene are reproduced, it is necessary to know something about the picture tube. The tube used in the basic color receiver is known as the tricolor picture tube. Three separate electron guns are incorporated in this type of tube in order to accommodate the three color signals. The coating on the face of the tube consists of phosphor dots, which are arranged in triangular groups of three. One dot in each trio emits red light, one emits green light, and one emits blue light.

When operating properly, the electron beam from the red gun activates only the red dots, the beam from the blue gun activates only the blue dots, and the beam from the green gun activates only the green dots. The phosphor dots are closely spaced so that when more than one dot in each trio is activated, the total light emission will blend to form one color. For example, when the light emissions from all three phosphors are equal, the screen appears white. If the red and green phosphors are activated equally, the resultant hue appears to be yellow. The blending into a single hue of the color dots on the picture-tube screen is based on the principle that the human eye cannot resolve the separate colors at normal viewing dis-

tance. As a result, the total light emitted from the dot combination appears as a single color.

The foregoing information should be helpful in understanding the requirements of the matrix section. A detailed discussion of the picture tube will be presented later.

In order to correctly produce the colors of an image, the matrix section must fulfill certain requirements. For instance, when a fully saturated red portion of the scene is to be reproduced, the amplitude of the video signal that is applied to the red gun must be at a maximum value, and the amplitude of the signals applied to the green and blue guns must be zero. During the time when a fully saturated green is to be reproduced, the amplitude of the signals applied to the red and blue guns must be zero and that of the signal applied to the green gun must be at a maximum value. The amplitude of the blue signal must be at a maximum value, and those of the red and green signals must be at zero when a fully saturated blue is being reproduced. If white is to be reproduced, the amplitude of all three of the color signals must be at maximum values, because white contains all colors.

During the transmission of a composite color signal, the content of each of the Y, I, and Q signal voltages at the transmitter is known (from Chapter 3) to be as follows:

$$E_Y = .30E_R + .59E_G + .11E_B \qquad (4)$$
$$E_I = .74 (E_R - E_Y) - .27 (E_B - E_Y) \qquad (8)$$
$$E_Q = .48 (E_R - E_Y) + .41 (E_B - E_Y) \qquad (9)$$

From these equations, it can be determined that:

$$E_R = .96E_I + .63E_Q + 1.00E_Y \qquad (13)$$
$$E_B = -1.11E_I + 1.72E_Q + 1.00E_Y \qquad (14)$$
$$E_G = -.28E_I - .64E_Q + 1.00E_Y \qquad (15)$$

E_R, E_B, and E_G represent the desired red, blue, and green signal voltages that are to be applied to the picture-tube guns. The mathematical proof for these equations is presented in footnote 1.

It can be seen that the voltage combinations needed to produce the color signals consist of plus and minus values of E_I and E_Q. The plus and minus signs designate the polarities of the signal voltages. For instance, the minus sign in the expression $-.28E_I$ indicates an I signal with a negative polarity. When no sign is used, the polarity is considered positive.

Let us suppose that the color-bar pattern shown at the top of Fig. 8-1 is used as the transmitted scene. If all of the colors are fully saturated, the relative amplitudes of the I, Q, and Y signals used in the makeup of the composite signal at the transmitter will be as shown below the bar pattern. The amplitude of the luminance signal during the reproduction of a white of full brightness is considered as the standard for unity. Therefore, E_Y is shown to have an amplitude of 1.00 during the transmission of the white bar.

The amplitudes of the luminance signal and the color-difference signals during the transmission of the color-bar pattern shown are based on this unity figure. For example, when the camera scans a fully saturated red, the amplitude of E_I is .60, the amplitude of E_Q is .21, and the amplitude of

---

1. The content of the I and Q signals is known to be:

$$E_I = .74 (E_R - E_Y) - .27 (E_B - E_Y) \qquad (8)$$
$$E_Q = .48 (E_R - E_Y) + .41 (E_B - E_Y) \qquad (9)$$

If both sides of equation 8 are multiplied by .41 and if both sides of equation 9 are multiplied by .27, then:

$$.41 E_I = .30 (E_R - E_Y) - .11 (E_B - E_Y)$$
$$.27 E_Q = .13 (E_R - E_Y) + .11 (E_B - E_Y)$$

Adding the foregoing equations, we arrive at:

$$.43 (E_R - E_Y) = .41E_I + .27E_Q$$

Divide by .43, and solve for E_R:

$$E_R - E_Y = .96E_I + .63E_Q$$
$$E_R = .96E_I + .63E_Q + 1.00E_Y \qquad (13)$$

By substituting this value of E_R in equation 8, we find:

$$E_I = .74 (.96E_I + .63E_Q) - .27 (E_B - E_Y)$$

From the foregoing equation, E_B may be found:

$$E_B = -1.11E_I + 1.72E_Q + 1.00E_Y \qquad (14)$$

It was shown in Chapter 3 that:

$$E_G - E_Y = -.51 (E_R - E_Y) - .19 (E_B - E_Y)$$

Substituting values of E_R and E_B from equations 13 and 14, it is found that:

$$E_G - E_Y = -.51 (.96E_I + .63E_Q)$$
$$\qquad -.19 (-1.11E_I + 1.72E_Q)$$

Solve the foregoing equation for E_G:

$$E_G - E_Y = -.49E_I - .32E_Q + .21E_I - .32E_Q$$
$$E_G = -.28E_I - .64E_Q + 1.00E_Y \qquad (15)$$

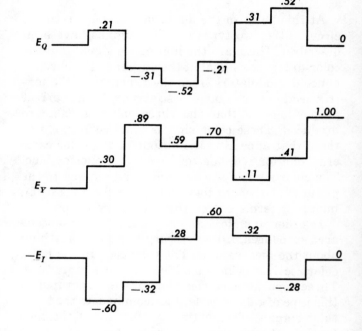
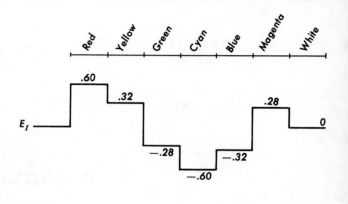
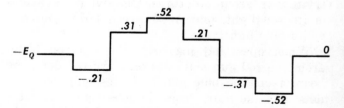

Fig. 8-1. Amplitudes of the video signals developed during a color-bar transmission.

E_Y is .30. If the color of the image is changed to green, the amplitude of E_I becomes $-.28$, the amplitude of E_Q becomes $-.52$, and the amplitude of E_Y becomes .59. If the polarity of the E_I or E_Q signal is inverted (as it is by an amplifier stage), the positive levels become negative and the negative levels become positive. The negative polarities of E_I and E_Q are also shown in Fig. 8-1 to provide a convenient reference for the reader.

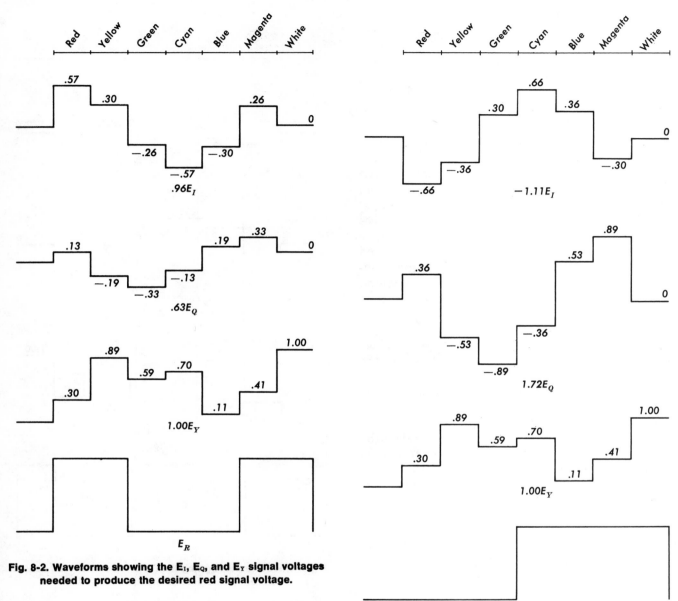

Fig. 8-2. Waveforms showing the E_I, E_Q, and E_Y signal voltages needed to produce the desired red signal voltage.

In accordance with equations 13, 14, and 15, certain proportions of the signals shown in Fig. 8-1 must be combined in the matrix to produce the three color signals. For instance, equation 13 states that .96 of the positive E_I signal, .63 of the positive E_Q signal, and 1.00 of the positive E_Y signal are required to produce the red signal. The amplitudes of the signals shown in Fig. 8-2 are of these proportions. During the transmission of the red bar, the amplitude of .96E_I is seen to be .57 (which is 96 percent of .60). The amplitude of 63E_Q at this time is .13 (which is 63 percent of .21). The amplitude of 1.00E_Y during the transmission of red is .30.

During the time the red bar is being scanned, the amplitudes of the I, Q, and Y voltages that

Fig. 8-3. Waveforms showing the E_I, E_Q, and E_Y signal voltages needed to produce the desired blue signal voltage.

are combined in the red matrix section have a ratio of .57 to .13 to .30, respectively. Since the addition of these amplitudes totals unity, the amplitude of the red signal is at maximum when red is being reproduced. If the same procedure is followed for all seven bars, it will be seen that the red signal will also have a maximum amplitude during the reproduction of yellow, magenta, and white. The combined amplitudes during the scanning time of green, cyan, or blue add up to zero; therefore, the red signal does not contribute to the reproduction of these colors.

The voltages determined by equation 14 are graphically shown in Fig. 8-3. These values are combined in the blue matrix section to produce the blue signal. The addition of amplitudes shows E_B to have a maximum amplitude for cyan, blue, magenta, and white; and the amplitude of E_B is zero for red, yellow, and green. This is the desired result, since E_B contributes to the colors that contain blue and does not contribute to the colors that are void of blue.

The values of the three signal voltages applied to the green matrix section were obtained from equation 15 and are shown in Fig. 8-4. The value of E_G is obtained by adding the values of these three signals. The resultant signal is shown to have maximum amplitude for yellow, green, cyan, and white. This indicates that the green signal contributes to the reproduction of each of these colors. E_G is at zero amplitude during the scanning time of red, blue, and magenta; therefore, the green signal does not contribute to these colors.

A comparison of the three color signals shows how the individual color bars are produced on the viewing screen. During the scanning time of the red bar, the red signal has a maximum amplitude and the blue and green signals have zero amplitude. Only the red phosphor is illuminated, and a saturated red is produced on the screen. When the yellow bar is being scanned, both the red and green signals have maximum amplitudes and the amplitude of the blue signal is zero. The red and green phosphors are both activated; however, the eye cannot see these colors separately. Instead, the total light emitted appears to be yellow. Cyan is produced when the light emissions from the blue and green phosphors are equal. When the light outputs from the red and blue phosphors are equal, the eye will see a fully saturated magenta.

The only time when all three color signals have maximum amplitudes is during the reproduction of white. This is to be expected since white contains all colors. It is interesting to note that during the scanning time of the white bar, the amplitudes of E_I and E_Q at the input of the matrix sections are at zero. The amplitude of the E_Y signal is at maximum in each case, which indicates that white is represented by the luminance signal only.

THEORY IN PRACTICAL APPLICATION

Now that the purpose of the matrix has been discussed, let us examine an actual circuit. The schematic in Fig. 8-5 shows the I and Q matrix circuit used in an early-model color receiver. Al-

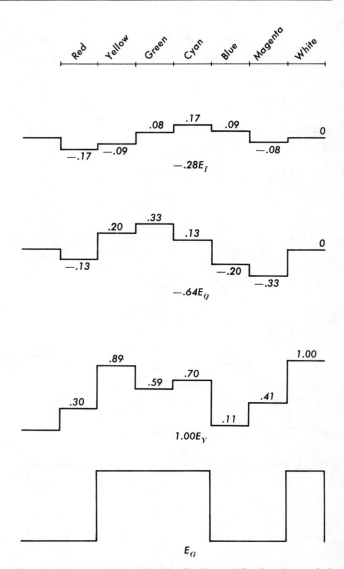

Fig. 8-4. Waveforms showing the E_I, E_Q, and E_Y signals needed to produce the desired green signal voltage.

though this circuit is now obsolete, it has been retained in this discussion because it clearly shows the relationship of the E_I, E_Q, and E_Y signals in the matrix section. Modern matrix circuits, which are somewhat simpler, are discussed later in this chapter.

The triode sections of two 6AN8 tubes are used as phase splitters. A Q signal having a negative polarity is available at the output of the Q demodulator and appears at the grid of V10B. An E_Q signal having a positive polarity is available at the plate of this stage, and an E_Q signal having a negative polarity is available at the cathode. Since the polarity of the I signal from the I demodulator has been reversed through the I amplifier stage, the signal at the grid of V32B has a positive po-

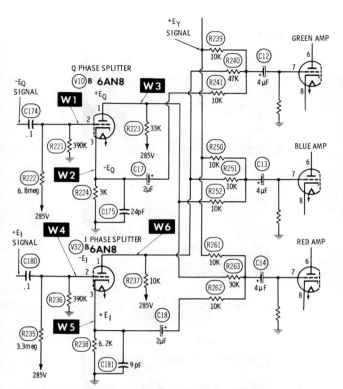

Fig. 8-5. Matrix circuit used in an early model color receiver.

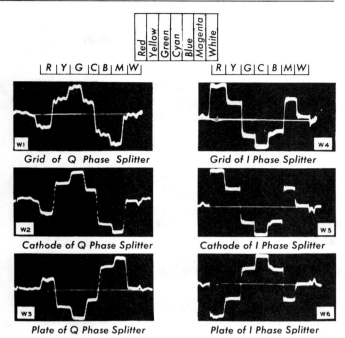

Fig. 8-6. Waveforms of the signal voltages present at the I and Q phase-splitter stages during the transmission of the color-bar pattern shown at the top.

larity. The polarity of E_I at the plate of this stage is negative, and its polarity at the cathode is positive. During a typical color-bar transmission, these signal voltages are like those shown by the waveforms in Fig. 8-6.

Refer to Fig. 8-5, and note that the positive E_Q signal is fed to the red and blue matrix networks and that the negative E_Q signal is applied to the green matrix. The negative E_I signal is fed to the green and blue matrix networks, and the positive E_I signal is applied to the red matrix. A positive E_Y signal from the output of the luminance channel is applied equally to all three matrix networks. Note that the signals applied to each matrix network have the proper polarity to produce the three color signals.

Let us see how the specific amplitudes of E_I, E_Q, and E_Y are obtained. The E_Y signal is applied to each of the matrix circuits through R239, R250, and R261. All three of these resistors are equal in value; therefore, if the grid resistance to ground at each amplifier stage is assumed to be the same, the amount of Y voltage to each of these stages will be the same. The amplitude of the chrominance signal at the output of the bandpass amplifier will have a fixed relationship with the amplitude of the luminance signal. This is because, in this receiver, the contrast adjustment that controls the

gain of the bandpass amplifier is ganged with the contrast adjustment that controls the gain of the luminance amplifier.

The amplitude of the chrominance signal that is applied to the synchronous detectors can be varied through the use of the color-saturation control. Since this control will vary the amplitude of the E_I and E_Q signals, it can be adjusted so that the amplitude of the negative E_Q signal at the input of the green amplifier has the desired relationship to the amplitude of the E_Y signal at this point (see Fig. 8-4). The amplitude of the positive E_Q signal is determined by the gain of V10B and by the value of load resistor R223. The value of R223 is such that the amplitude of the positive E_Q signal at the input of the blue amplifier has the desired relationship to the amplitude of the E_Y signal (see Fig. 8-3).

The gain control in the cathode circuit of the I amplifier (not shown) can then be adjusted for the desired amplitude of the positive E_I signal at the input of the red amplifier. The amplitude of the negative E_I signal at the plate of the I phase splitter will then depend on the gain of V32B and the value of the plate-load resistor R237. The value of R237 is such that the amplitude of the negative E_I signal has the desired amplitude at the input of the blue amplifier.

A certain amount of the negative E_I signal is also required at the input of the green amplifier;

however, the amplitude of the signal at the plate of the I phase splitter is more than four times the desired amount. The 47,000-ohm resistor R240 and the resistance from the grid of the green amplifier to ground form a voltage-divider network. The same is true for the 10,000-ohm resistor R251 and the resistance from the grid of the blue amplifier to ground. It was originally assumed that the resistance from grid to ground was equal at each amplifier stage; consequently, the amount of negative E_I signal developed across the grid resistance of the green amplifier will be much less than that developed at the input of the blue amplifier. The value of R240 is such that the voltage division would produce the correct amount of negative E_I at the input of the green amplifier.

The proper voltages needed to produce the blue and green signals have been developed. A specified amount of positive E_Q voltage at the red amplifier will complete the voltages needed to produce the red signal. Since the amplitude of the positive E_Q signal at the plate of V10B is considerably greater than the amount needed at the input of the red amplifier ($1.72E_Q$ was required at the input of the blue amplifier), the voltage-division principle is again applied. The 30K resistor R263 causes a voltage division in the proper ratio to produce the desired amount of E_Q signal at the input of the red amplifier.

In actual practice, proper matrixing in this receiver can be accomplished by adjusting the hue, contrast, color-saturation, and I-gain controls. The grid resistance at each of the amplifier stages is a common load for the three signals applied to each matrix circuit. For this reason, a change in one signal amplitude will affect the amplitude of the other two signals across the same load circuit. As a result, the aforementioned controls must be adjusted alternately in order to obtain the desired signal ratios at the input of the individual matrix circuits.

Following this procedure, an NTSC color-bar signal is fed into the receiver and the controls are adjusted while the results are being observed on an oscilloscope. When the blue signal is balanced, as shown by waveform W7 in Fig. 8-7, the red and green signals should then appear as shown in waveforms W8 and W9. These two signals will automatically balance when the blue signal is correctly adjusted, since a prearranged relationship exists between the three matrix circuits. It can be seen that these three waveforms at the input of the matrix amplifiers conform to the requirements for proper color reproduction.

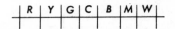

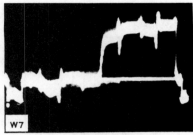

Grid of Blue Amplifier

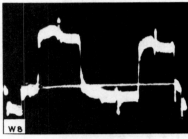

Grid of Red Amplifier

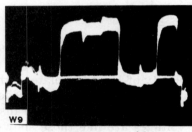

Grid of Green Amplifier

Fig. 8-7. Waveforms of the color-signal voltages at the input of the matrix amplifiers when the circuits are correctly balanced.

X AND Z MATRIXING

In the X and Z matrix shown in Fig. 8-8, the picture tube is used to complete the matrixing of the color signal. Simplification of the matrix circuitry as compared with that used in Fig. 8-5 is quite apparent. Included in this schematic is the circuitry ahead of the picture tube cathodes and, although it is not directly a part of the matrix, it does affect the color reproduction on the screen.

Adjustment of the green and blue drive controls will set a proper balance of signals to each of the three color guns to achieve a good black-and-white picture over the full-range from black through gray to white.

The X and Z signals, developed by the color demodulators, are applied to the grids of the $R - Y$ amplifier and the $B - Y$ amplifier. The X and Z signals contain all of the information needed to derive the three color signals $R - Y$, $B - Y$, and

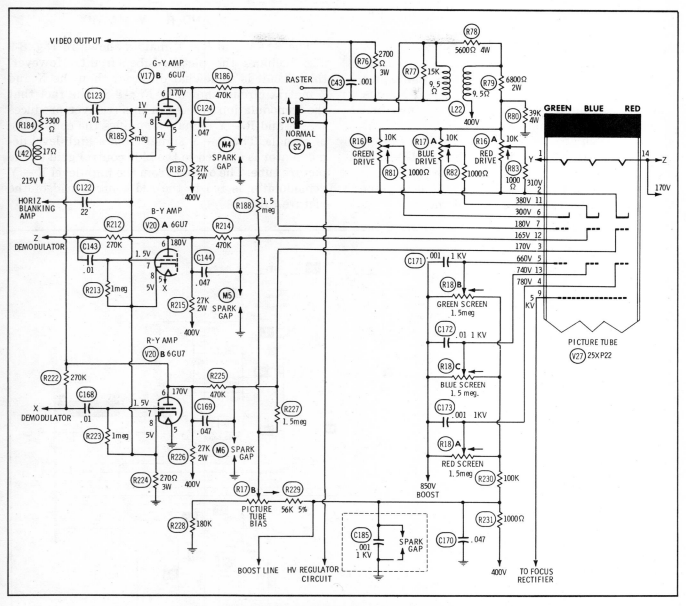

Fig. 8-8. X and Z matrix used in a tube-type color receiver.

G − Y. Although the X and Z signals go directly to the inputs of color-difference amplifiers, these stages do not directly amplify these individual signals. The common cathode resistor R224 becomes a signal source for all three amplifiers. The G − Y amplifier has a grounded grid, and the G − Y output will be an amplified version of the signal developed across the common cathode resistor R224. The signal produced across this resistor will always be a G − Y signal.

The G − Y signal is produced by the addition of the current through V20A and V20B. The application of signals X and Z to the grids of these tubes produces a resultant current through R224

which is equal to a G − Y signal. If R224 is considered a generator, then G − Y can be subtracted from the Z signal at the input of the B − Y amplifier and from the X signal at the input to the R − Y amplifier. The B − Y signal at the plate of the B − Y amplifier will then be equal to the Z signal minus the G − Y signal, which has been inverted 180 degrees and amplified by the 6GU7 triode. Note that the amplitude of the signals at the output of the matrix amplifiers is adequate to drive the grids of the picture tube.

The R − Y amplifier operation is identical to that of the B − Y amplifier, except that the signal is taken off at the junction of R225 and R227.

This reduces the amplitude of the plate signal by a factor of .7 and provides an amplitude-corrected signal from the R − Y amplifier.

The red, green, and blue signals are produced by subtracting the −Y, or negative luminance, signal from the R − Y, G − Y, and B − Y signals. This subtraction takes place in the picture tube by placing the Y signal on the cathode. The cathode signal is subtracted from the grid in a tube or amplifier, and the result in the case of G − Y is as follows:

(G − Y) − (−Y) = G − Y + Y = G, or a green signal, is produced on the screen of the crt.

B − Y AND R − Y MATRIX

The B − Y and R − Y matrix shown in Fig. 8-9 also includes the picture-tube circuit. However, this circuit is somewhat different than the X and Z matrix circuit shown in Fig. 8-8. The fact that the demodulation process in this receiver produces B − Y and R − Y signals simplifies the combining of signals. The B − Y signal from a high-level demodulator is applied to the blue control grid of the picture tube. The output from the high-level R − Y demodulator is fed to the red control grid of the picture tube.

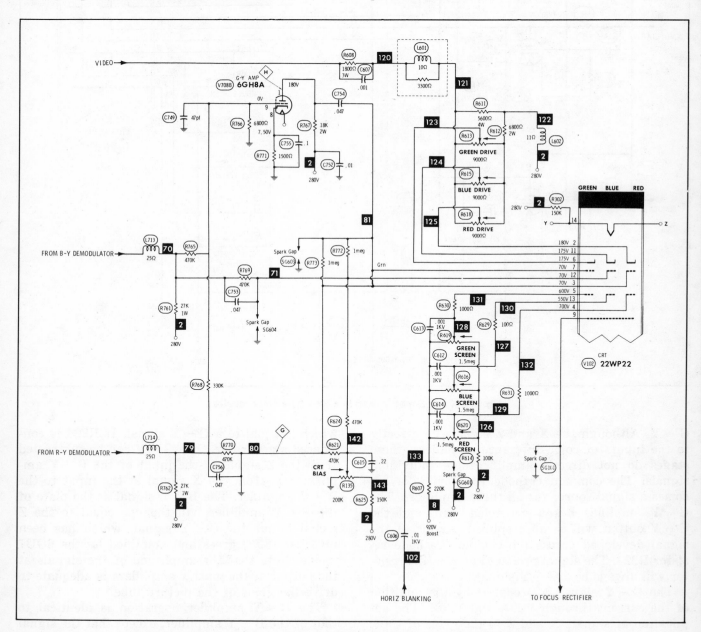

Fig. 8-9. A B − Y and R − Y matrix circuit used in a tube-type color receiver.

It has been shown previously that the $G - Y$ signal can be derived by combining $-.5(E_R - E_Y)$ and $-.19(E_B - E_Y)$. In this circuit, part of the $B - Y$ signal is fed to the grid of the $G - Y$ amplifier, V708B, through resistor R765. In turn, a portion of the $R - Y$ signal is also fed to the grid of the $G - Y$ amplifier. Since both the $B - Y$ signal and the $R - Y$ signal are positive, a negative $G - Y$ signal is developed across grid resistor R766. This signal is inverted and amplified to produce a positive $G - Y$ signal at the plate of V708B. This signal is then fed to the green control grid of the picture tube.

Like the X and Z matrix circuit shown in Fig. 8-8, the video signal in this circuit is fed to the green, blue, and red cathodes of the picture tube. In this manner, the green, blue, and red signals are developed at the picture tube. It should be remembered that the $B - Y$ signal can be developed by combining the $G - Y$ and $R - Y$ signals, or the $R - Y$ signal can be developed by combining proper portions of the $B - Y$ and $G - Y$ signals. Therefore, a circuit similar to the one shown in Fig. 8-9 can be used with $G - Y$ and $R - Y$ signals or with $B - Y$ and $G - Y$ signals as long as the two demodulated signals are combined in the proper proportions.

SOLID-STATE MATRIX CIRCUIT

The matrix circuit shown in Fig. 5-10 is from a solid-state color receiver. In this receiver the three color-difference signals are demodulated separately. Most solid-state color receivers employ separate demodulators for each of the three primary colors. This is particularly true for demodulator circuits using an IC. The demodulation of all three colors also simplifies the matrix circuit.

The outputs from the three demodulators are fed to their respective color-difference amplifiers. Since the color-difference amplifiers are connected as emitter-followers, there is no inversion of the color-difference signals at this stage. Each of the color-difference amplifiers is connected in series with the emitter circuit of its respective video output stage. The video signal is applied to the base of each of the three video output stages. In this manner, the color-difference signals are combined with the video signal at the video output stages to develop the green, blue, and red output signals. These signals, in turn, are coupled to the cathodes of the picture tube.

The emitter bias for each of the video output stages is determined by the adjustment of its respective background control. The adjustment of the green, blue, and red drive controls sets the proper balance between the color-difference signal and the video signal for each of the video output stages. The six controls are adjusted to produce a good black-and-white picture at all brightness levels.

DC RESTORATION

Unless direct coupling is used, the dc voltage levels of the three color signals is eliminated by the coupling capacitors between the various amplifier stages. If the correct dc levels are not restored, these signals will not accurately reproduce the background illumination. Instead, the background illumination will be represented by the average amount of voltage of the ac signals applied to the picture tube.

The waveform shown in Fig. 8-11A is representative of a scene which has a high value of brightness. The brightness control has been adjusted so that picture-tube cutoff is just above the blanking level. If a signal representing a scene of low brightness were to be transmitted, a condition like that shown in Fig. 8-11B would exist. The average bias on the picture tube does not change; consequently, the background illumination is still the same. In addition, the blanking level will not reach the cutoff level of the picture tube, and the retrace lines may be seen in the picture.

The desired dc level of the signal at the grid of the picture tube is illustrated by the waveform in Fig. 8-12. The sync tips are shown to be clamped at a predetermined dc level so that cutoff is achieved by the blanking pedestal regardless of the amplitude of the signal. It can be seen that the average voltage at the grid of the picture tube will be equal to the dc clamping level plus the average voltage produced by the ac signal.

One method of obtaining the foregoing condition is through the use of the circuit shown in Fig. 8-13. This circuit is used to restore the dc level to the red signal. The restoration circuits for the blue and green signals are identical to the one shown. The chroma signal is RC-coupled through C40 from the plate of the $R - Y$ amplifier, V4B, to the crt grid. The voltage on the crt grid is determined by the setting of the crt bias control, R78. At the end of each horizontal scan line, a negative pulse at the plate of the horizontal blanking amplifier is fed to the cathode of dc restorer diode CR7 through R78 and R24. This negative pulse causes the diode to conduct, and the crt grid is

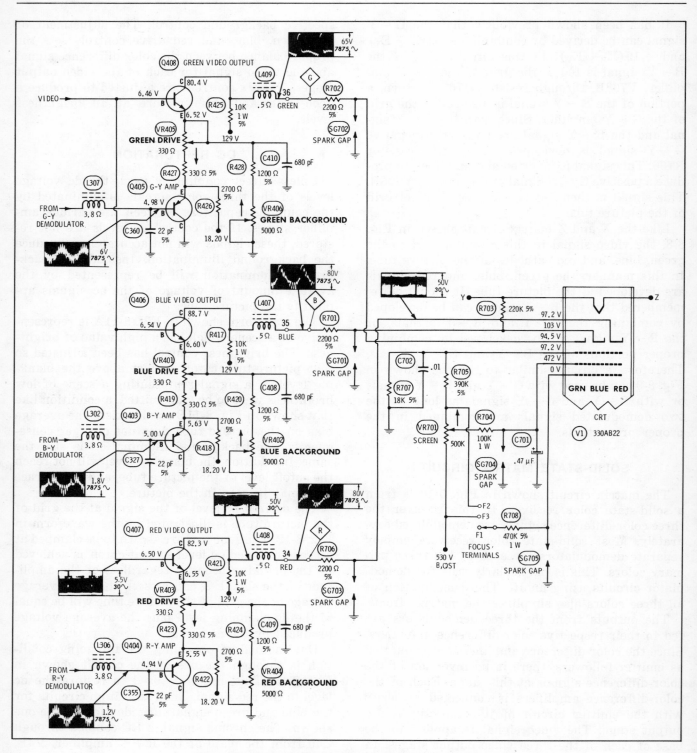

Fig. 8-10. Matrix circuit used in a solid-state color receiver.

clamped to the voltage on the wiper arm of the crt bias control.

Let us suppose that the signal representing a bright scene in Fig. 8-12A is present at the plate of V4B. During the negative excursions of the signal, the voltage on both sides of C40 will decrease by 60 volts. At this time, the horizontal blanking period occurs and the negative pulse on the cathode of CR7 causes the diode to conduct. As a result, the crt grid voltage returns to the

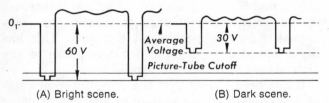

(A) Bright scene.   (B) Dark scene.

Fig. 8-11. Loss of dc level in the video signal.

value determined by the setting of the crt bias control.

When the signal at the plate of V4B goes in a positive direction, C40 will try to charge to a higher value. However, the charging time of C40 is considerably long due to the high value of R56 and is much greater than the scanning time of a single horizontal line. Therefore, the charge across C40 remains fairly constant until the brightness level of the scene changes. The voltage on the crt side of the capacitor will follow the signal variations at the plate of V4B. Consequently, the voltage at the red crt grid will have a value equal to the signal plus the voltage at the cathode of CR7. The negative peaks of the signal are clamped to the dc level determined by the setting of the crt bias control. Note that this condition conforms with that shown in Fig. 8-12A.

Now let us see what happens when a signal representing a dark scene is being received. During the negative excursions of the signal, the voltage on both sides of C40 will decrease by 30 volts. Since the horizontal-sync pulse occurs at the most negative portion of the signal, diode CR7 is again biased into conduction by the negative pulse from the horizontal blanking amplifier. Again, the crt grid voltage returns to the value determined by the setting of the crt bias control.

When the signal goes in the positive direction, C40 tries to charge to a higher level but does not have time to do so. (The signal is positive for approximately 60 microseconds, whereas the time constant of C40 and R56 is 22,000 microseconds.) When the signal at the plate of V4B traverses from the negative peak to the positive peak, the voltage at the crt side of C40 will increase by the

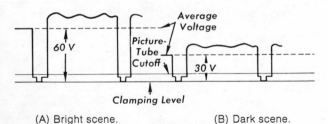

(A) Bright scene.   (B) Dark scene.

Fig. 8-12. Video signals with dc restored.

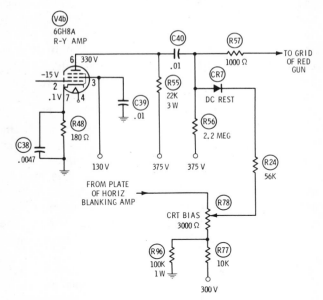

Fig. 8-13. A dc restorer circuit used in a late-model, tube-type color receiver.

same amount. As a result, the negative peaks of the signal at the anode of the diode are clamped at a specific dc level, as shown in Fig. 8-12B.

Because of the conduction of CR7 during the horizontal blanking period, the negative peak of the signal on the crt side of C40 is clamped to the same dc level, regardless of the amplitude of the signal. As mentioned previously, a dc restorer circuit is employed in the grid circuit of each crt gun. The cathode of each of the dc restorer diodes is connected to the crt bias control through a 56K resistor.

SIGNAL-BIAS DC RESTORATION

Shown in Fig. 8-14 is the B − Y amplifier used in another tube-type color chassis. Note that ap-

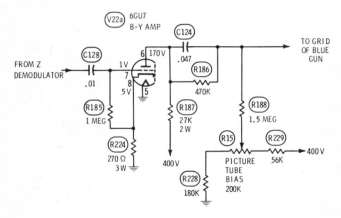

Fig. 8-14. B − Y amplifier.

parently there is no provision for dc level restoration. In this circuit the grid of the B − Y amplifier is signal-biased, as are the grids of the G − Y and R − Y amplifiers. The grid of V22A in Fig. 8-14 has a 1-megohm dc path to the cathode. Capacitor C123 will charge when a positive-going signal pulse exceeds the bias of the amplifier and will discharge through the one-megohm resistor. The voltage drop across R185 will provide the bias for the B − Y amplifier, and this bias will vary with the maximum amplitude of the color signal.

It must be remembered that the signals present at the input of these matrix amplifiers contain only the color information and that the sync pulses were removed at the chrominance amplifier. The bias variations then are affected only by the amplitude of the color information.

When the color receiver is reproducing a black-and-white signal, the color circuits are not functioning. The G − Y, B − Y, and R − Y amplifiers conduct at an average level, and the bias on the picture-tube grids remains constant. Bias for the picture tube is set by adjusting the picture-tube bias control, R15. This control sets the normal bias level for all three picture-tube guns. This voltage is the level to which the color signal will be clamped.

During a color transmission, the change in signal bias on the grid of V22A in Fig. 8-14 is equal to the peak input signal to the stage. At the plate, a reference voltage of 170 volts has been estab-lished, and the clamp bias on the grid will increase the plate voltage by an amount equal to the peak-to-peak voltage appearing at the plate. This has the effect of clamping the negative peaks of the plate signal at a 240-volt level.

It has been shown that the color-difference signals produced at the output of the demodulator circuit can be combined with the luminance signal so that three color signals will be formed. Each of these color signals represents one of the three primary colors contained in the original scene.

The construction and function of the color picture tube and the circuits and components associated with this tube are explained in detail in the next chapter.

QUESTIONS

1. What is the output of a demodulator during the time a white bar is being reproduced?
2. When the luminance signal is applied to the cathode of the picture tube, what color signals are applied to the grids?
3. What is the polarity of the luminance signal applied to the cathodes of the color picture tube?
4. What purpose is served by the dc restorer?
5. Can the picture tube be used as a part of the matrix?
6. Are diodes always required in a color receiver to provide dc restoration?

Chapter 9

The Color Picture Tube and Associated Circuits

The color picture tube that has been developed for use in a color receiver is considerably different from the picture tube used in a monochrome receiver. The color picture tube must be capable of producing an image in accordance with the brightness and color variations of the televised scene. The monochrome picture tube has only one duty to perform, and that is to reproduce the scene in accordance with the brightness variations of the scene.

The monochrome and the color picture tubes, however, do have some basic characteristics that are common to both types. In outside appearance they are very similar—both are conical-shaped vacuum tubes and both have a screen on which the picture is formed and a gun assembly from which the electrons originate.

CHARACTERISTICS OF THE THREE-BEAM PICTURE TUBE

Some of the characteristics of the three-beam picture tube are: it has a phosphor-dot screen made up of three different phosphors; it has three beams originating from three electron guns to energize each of the different phosphors; and it has a shadow mask to allow each beam to strike only the correct set of phosphor dots. There are, therefore, three major parts in a color picture tube. These are a phosphor viewing screen, a shadow or aperture mask, and an electron-gun assembly. An illustration showing the location of these parts appears in Fig. 9-1.

Let us investigate the characteristics of the three-beam color picture tube by first discussing the three major parts.

The Phosphor Screen

The screen of the monochrome picture tube is made up of a mixture of phosphorescent material which, when energized by an electron beam, will emit white light. This material is placed on the face plate of the tube in the form of a solid screen. The viewing screen of a color picture tube is also made up of phosphorescent material, but since three different phosphors are used, the screen of the color picture tube is different from that of the monochrome tube. The phosphors are the type that will emit colored light when they are energized by electrons, because color has to be reproduced on the screen. Since three additive primaries are employed in color television, three different phosphors are used. The phosphors are deposited on the viewing surface in the form of dots in a set pattern.

In conventional color picture tubes, the three color phosphors are deposited on the face plate in a pattern of dots. Some newer color picture tubes, however, have the three color phosphors arranged in thin vertical stripes rather than dots. These newer tubes will be covered later in this chapter. First, we will discuss the conventional color picture tubes that employ the phosphor-dot arrangement.

These dots are placed very close together, but they do not overlap or touch each other. A third of all the phosphor dots emit red light, another third of them emit green light, and the other third emit blue light. The drawing shown in Fig. 9-2 represents a magnified portion of the viewing screen. This shows the arrangement of the dots as they are viewed from the front of the tube.

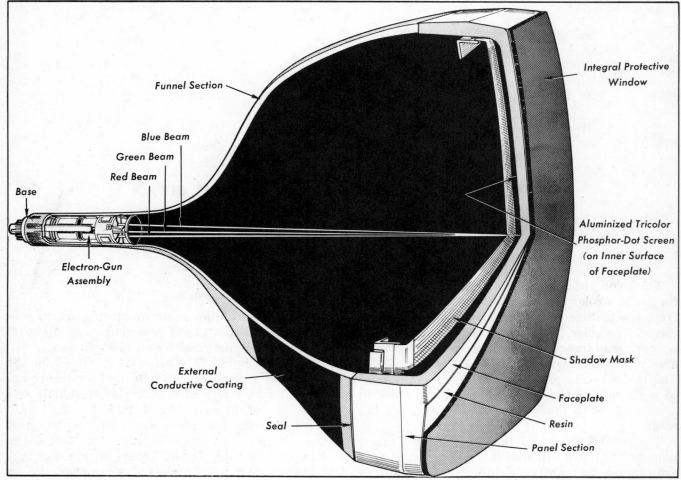

Funnel Section

Blue Beam

Green Beam

Red Beam

Base

Electron-Gun
Assembly

External
Conductive Coating

Seal

Integral Protective
Window

Aluminized Tricolor
Phosphor-Dot Screen
(on Inner Surface
of Faceplate)

Shadow Mask

Faceplate

Resin

Panel Section

Fig. 9-1. Details of color picture-tube construction showing location of major parts.

Note that the dots are arranged into a definite pattern, which is in the form of trios (hereafter called triads) with a red-phosphor dot, a green-phosphor dot, and a blue-phosphor dot forming one triad. Fig. 9-3 is a drawing of a portion of the screen as viewed from the front of the tube, and it illustrates the arrangement of the triads. This pattern appears over the entire screen.

When an electron beam strikes a red-phosphor dot in a triad, that dot will glow with a red light. When the beam strikes a green-phosphor dot, that

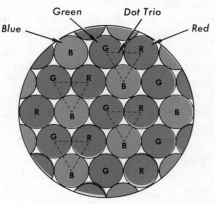

Fig. 9-2. Arrangement of the phosphor dots on the screen of the color picture tube.

Fig. 9-3. A portion of the phosphor-dot screen showing the arrangement of the dot trios.

dot will glow with a green light. The blue-phosphor dot will glow with a blue light when it is energized with a beam. The characteristics of the human eye are such that the light emissions from the three phosphors cannot be distinguished separately at normal viewing distance. Instead, the eye blends the light from the three sources to give the appearance of a single color. For example, when the light outputs of all three phosphors are equal, each dot will glow with its respective color, but the eye blends the three lights together so that the screen will appear to be white. By controlling the energization of the phosphors, it is possible to produce a variety of colors corresponding to the hues in the visible light spectrum. For instance, when only the red and the green phosphors are energized, the two light sources are blended together by the eye, and the color yellow is seen. If the green and the blue phosphors are energized, the eye sees the color cyan.

Shadow or Aperture Mask

It has been stated that the color picture tube has three electron beams. One beam is used for energizing the red-phosphor dots, one for the green-phosphor dots, and the other for the blue-phosphor dots. These three beams must be made to strike their respective set of dots at all times. To make this possible, a shadow mask is placed in the path of the electron beams, directly behind the phosphor screen.

A drawing illustrating the structure of the shadow mask appears in Fig. 9-4. The shadow mask consists of a thin sheet of metal that has been etched with a series of very small holes by a photoengraving process. This mask is made large enough to cover the entire phosphor screen. There are as many holes in the mask as there are triads on the phosphor screen—one hole for each dot

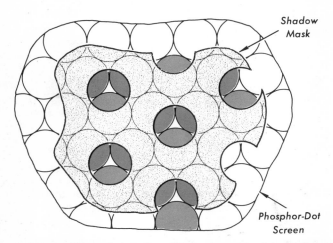

Fig. 9-5. View of phosphor screen of a color picture tube as seen through the aperture mask.

triad. The placement of the mask in respect to the phosphor dots is shown in Fig. 9-5. A red, green, and blue dot can be seen through each hole in the mask. Fig. 9-6 shows how electron beams from separate sources can be directed through a single hole in such a way that each beam will energize a separate dot on the screen. The displacement marked D', which is between any two points on the phosphor screen, is directly proportional to the displacement marked D, which is the distance between the two electron-beam sources. If displacement D increases, displacement D' increases a proportional amount. Since the mask is placed very close to the screen, displacement D' is very much smaller than displacement D. Displacement D can be in any direction within a plane that is perpendicular to the paper at the line that is indicated as the plane of electron sources. Since the positions of the electron guns in the tube are fixed,

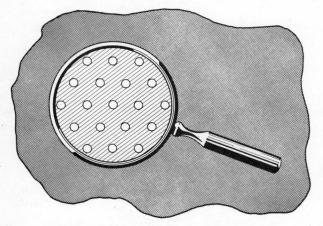

Fig. 9-4. Pattern of holes in aperture mask.

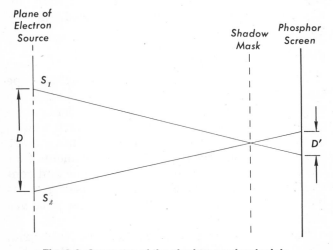

Fig. 9-6. Geometry of the shadow-mask principle.

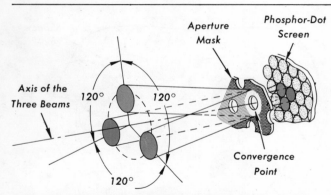

Fig. 9-7. Relationship between electron beams, aperture mask, and phosphor screen in color picture tube.

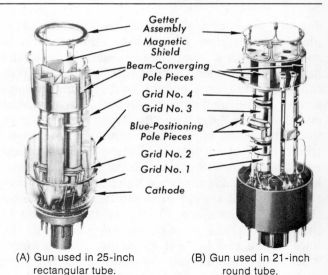

(A) Gun used in 25-inch rectangular tube.

(B) Gun used in 21-inch round tube.

Fig. 9-8. Electron-gun assembly.

the exact placement of the phosphor dots can be determined.

Fig. 9-7 shows the relationship of the electron beams, the aperture mask, and the phosphor-dot screen. The blue beam is shown as originating from the source on the top, the red beam from the source on the lower right, and the green beam from the source on the lower left. The three beams are controlled in such a way that they converge and diverge at the same holes in the aperture mask as they are scanned across the screen, and therefore each beam strikes only its respective set of color dots. The blue beam hits the blue-phosphor dot of the particular triad indicated in Fig. 9-7, and the red and green beams hit their respective dots in this triad. This triad of dots can be likened to the spot produced on a monochrome tube as the electron beam strikes the phosphor screen. Just as the brightness of this spot can be controlled in the monochrome tube by varying the intensity of the beam, so can the brightness of the triad in the color tube be changed by controlling the total intensity of the three beams. In addition, however, the beams in a color tube can be controlled individually, and this makes possible the reproduction of any desired hue.

Electron-Gun Assembly

As was stated previously, the color picture type employs three electron guns. Each is a complete unit in itself, and all three guns are identical in physical appearance and operation. Each gun is similar to the one used in a monochrome picture tube. The differences lie in the extra elements that have been added to achieve proper focus and control of the beams. The electron guns used in a 21-inch round tube and 25-inch rectangular tube are shown in Fig. 9-8.

Each gun contains a heater, cathode, No. 1 grid, No. 2 grid, No. 3 grid, and No. 4 grid. Grids No.

1 and No. 2 serve the same purpose as they do in a monochrome picture tube, No. 1 being the control grid and No. 2 the accelerating anode. Grid No. 3 in the color tube is the focus electrode. The focus electrodes of the three guns are electrically connected. Grid No. 4 serves as the final anode. The high-voltage anode of the color picture tube consists of grid No. 4 and the inside coating to which the aperture mask and the phosphor screen are connected.

An end-view drawing of the gun assembly showing the placement of each gun with respect to the axis of the gun assembly appears in Fig. 9-9. Note that the three guns are spaced symmetrically around the central axis of the gun assembly. The gun assembly is held in place by glass supports.

Requirements for Control of the Beams

Since there are three beams in the color picture tube, the requirements for beam control in this

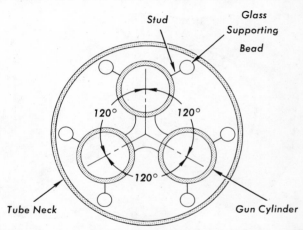

Fig. 9-9. End view of the gun assembly in a color picture tube.

tube are more stringent than they are in the monochrome picture tube. In the latter, we are concerned with only one controlling operation (other than deflection). This operation is the proper focusing of the beam throughout the scanning of the raster. The color picture tube has two additional requirements for controlling the beams. First, the beams must be so aligned that they will strike only their respective color-phosphor dots; and second, the beams must be made to pass simultaneously through the same holes in the aperture mask. To accomplish these requirements, precise control of the beams must be maintained.

After the electrons are emitted from the cathode of each gun, the intensity of the beam is governed by the action of the control grid of that gun. Then each beam is accelerated by the action of grid No. 2 (the acceleration anode). After the beams pass the point of acceleration, they are acted upon in such a way that they will converge at the aperture mask.

The three beams must be in perfect alignment with each other and with the central axis of the tube before they pass through the convergence field. Because of production tolerances, the electron guns may not be in perfect alignment with each other; the gun assembly may be turned slightly around the central axis of the tube with respect to the viewing screen, and the central axis of the gun assembly may not be parallel with or may not exactly coincide with the central axis of the picture tube. Fig. 9-10 illustrates these misalignments. Corrective measures must be taken to compensate for unavoidable variations that occur in the manufacturing process. If corrective measures were not taken, the three beams would not travel down the neck of the tube in the correct positions.

First of all, the three beams are aligned so that they are equidistant from each other and from the central axis of the gun structure. In order to obtain this result, each beam is acted upon by a separate beam-positioning magnet which is mounted outside the neck of the tube. By proper adjustment of the three magnets (called static convergence magnets), each beam is positioned with respect to the other two beams.

With the three beams correctly aligned with respect to each other, the entire system of beams has to be oriented with respect to the central axis of the tube. This is accomplished by the purity magnet which is placed around the neck of the tube and which affects all three beams equally. The adjustment of this magnet will move all three beams the same amount until they are properly aligned with the central axis of the tube.

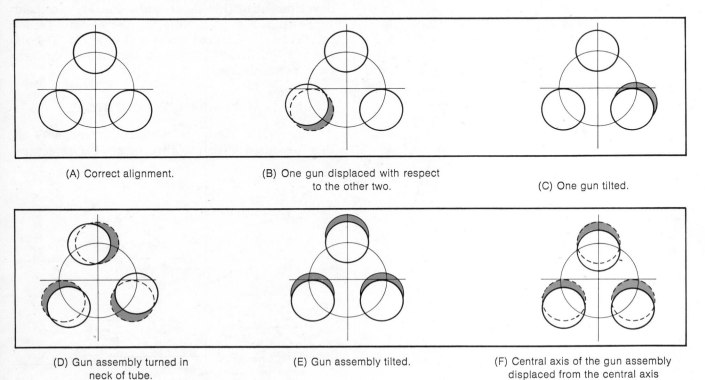

(A) Correct alignment.

(B) One gun displaced with respect to the other two.

(C) One gun tilted.

(D) Gun assembly turned in neck of tube.

(E) Gun assembly tilted.

(F) Central axis of the gun assembly displaced from the central axis of the tube.

Fig. 9-10. Drawings illustrating misalignment of the electron gun. The dashed circles indicate the correct position of the guns.

The three beams are brought into focus by the action of the No. 3 grids. These grids are electrically connected, which means that all three grids will have the same potential with respect to ground.

After being focused, the beams enter a convergence field which causes them to cross over at the aperture mask and strike the dots of the correct color. In most types of tubes now being used, this convergence field is obtained through the use of electromagnets that are mounted around the neck of the tube.

Up to this point, beam control has been considered as though the beams were undeflected or in a static state. However, the beams are constantly being deflected across the face of the tube. When they are deflected toward the edges of the raster, they must converge at the aperture mask the same as they do at the center of the mask. It can be seen from the drawing in Fig. 9-11 that this would not happen if special convergence forces were not employed. The control of the beams while they are in a changing state is referred to as dynamic convergence.

In Fig. 9-11 it can be seen that the beams converge farther and father behind the aperture mask as the deflection angle is increased. This is because the distance from the center of deflection to the mask increases as the angle of deflection is increased. This distance changes constantly in a set pattern. Since the horizontal- and vertical-scanning rates determine the rate of change in the deflection angles of the beams, energy from the horizontal- and vertical-deflection circuits can be used to obtain dynamic convergence. As the deflection angle is increased, the convergence force is decreased, therefore allowing the beams to converge at a point farther from the deflection plane. Dynamic convergence is obtained by varying the strength of the convergence field at the horizontal- and vertical-scanning rates.

The control of the beams in the color picture tube must be much more precise than the control of the beam in the monochrome picture tube. Stray magnetic fields within close proximity of the picture tube affect the beams. It was found that even the relatively weak magnetic field of the earth has an effect on the beams of a color picture tube. In order to counteract the effect of these magnetic fields, many color picture tubes are shielded.

The method of providing beam convergence in early color television receivers was to use grid No. 4 in the electron gun assembly as a convergence element. With the exception of grid No. 4, these guns were similar to the electron guns used in modern color picture tubes that employ magnetic convergence. In these older tubes, an electrostatic convergence field was developed between grid No. 4 and the coating on the inside of the tube neck. By controlling the voltages applied between grid No. 4 and the neck coating, it is possible to obtain convergence over the entire raster. The development of this electrostatic convergence field is shown in Fig. 9-12. Note the electrostatic focusing field between grid No. 3 and grid No. 4. This method is still used to obtain focusing in modern color picture tubes.

AUXILIARY COMPONENTS

This section concerns auxiliary components as they are used with the three-beam color picture

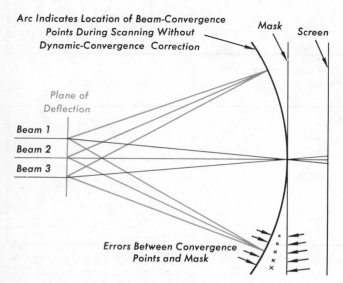

Fig. 9-11. Drawing illustrating the necessity for dynamic convergence.

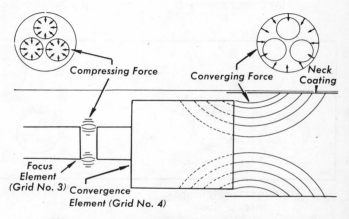

Fig. 9-12. Location of the electrostatic fields at each end of the convergence element.

tube employing the magnetic method of convergence.

As pointed out in the first part of this chapter, an exacting relationship between the electron beams and the mask and screen must be maintained to obtain optimum performance from the three-beam tube. Even though close tolerances are observed in the manufacturing process, some variations exist. One such condition will result if the guns are not positioned properly in the gun assembly. A second variation exists when the axis of the gun assembly does not coincide with the central axis of the tube. Another kind of variation is caused when the gun assembly is turned slightly in the neck of the tube.

In order to minimize the errors that are produced by these variations, it is necessary to use certain auxiliary corrective devices to control the

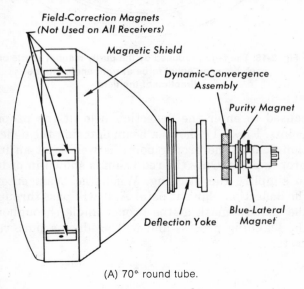

(A) 70° round tube.

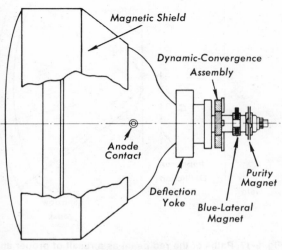

(B) 90° rectangular tube.

Fig. 9-13. Auxiliary components used on color picture tubes.

position of each beam with respect to the shadow mask and viewing screen. Other auxiliary components are used as a precaution against the effects of stray magnetic fields on the electron beams. Fig. 9-13A shows the auxiliary components used with a 70-degree three-beam tube employing magnetic convergence. In the order of their positions starting from the tube base, the external components are the blue-lateral correction magnet, the purity magnet, the beam-positioning magnets, the convergence electromagnets, and the deflection-yoke assembly.

Fig. 9-13B shows the auxiliary components on the 90-degree, rectangular color-picture tubes. Notice that the purity magnet is located nearest the picture-tube socket. It is positioned over the cathodes of the three guns.

Color purity is achieved when each beam strikes only its respective set of color dots. The adjustment of the purity magnet, the position of the deflection yoke, and proper degaussing play important parts in obtaining color purity.

The beam-positioning magnets and the blue-lateral correction magnet are used in conjunction with the convergence electromagnets to obtain beam convergence at all points on the shadow mask. If the proper adjustments are made to the beam-positioning magnets and to the blue-lateral correction magnet, and if the proper voltages are applied to the convergence electromagnets, the points at which the three beams strike the shadow mask will coincide throughout the scanning process. Under these conditions, optimum convergence is obtained.

The Color-Purity Magnet

The photograph in Fig. 9-14 shows that the purity magnet is similar to the centering device used with most types of monochrome picture tubes. Actually, this device uses two rings that are composed of magnetic materials. The arrangement of the molecules in each ring is such that one half of each ring is a north pole and the opposite half is a south pole. The alignment of the molecules in either ring can be seen in Fig. 9-15.

Each of the magnetic rings in the color-purity device may be rotated 360 degrees. In this manner, the unlike poles of each ring may be placed adjacent to each other so that no appreciable field will exist in the space at the center of the rings. If one magnet is rotated slightly, a weak magnetic field is produced. This field becomes increasingly stronger as one ring is rotated with respect to the other, and the field reaches maximum strength

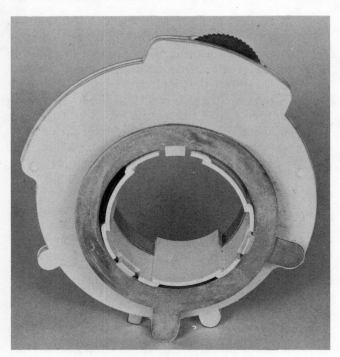

Fig. 9-14. Color purity magnet.

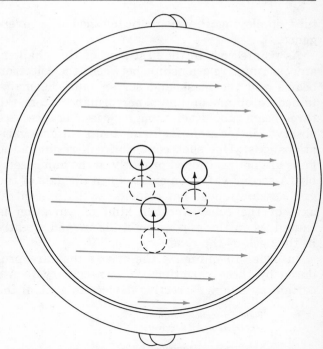

Fig. 9-16. The field produced by the purity magnet exerts on each beam an equal force at right angles to the direction of the field.

when the like poles of the two rings are adjacent to each other.

Fig. 9-16 shows the magnetic field produced in the space at the center of the purity device. This field is fairly uniform and exerts an equal force on all three beams. The force is at right angles to the direction of the magnetic field.

With proper adjustment of the strength and direction of the purity field, the beams can be

caused to enter the deflection field at the proper points. Fig. 9-17 shows a beam entering the deflection field at the proper point as well as at an improper point. Only the red beam is shown in order to simplify the drawing. When the beam enters the deflection plane at point A, it will pass through the shadow mask at the proper angle throughout the scanning process and will produce a pure red raster.

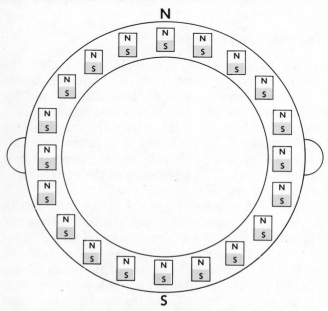

Fig. 9-15. Alignment of the molecules in the rings used in the color purity magnet.

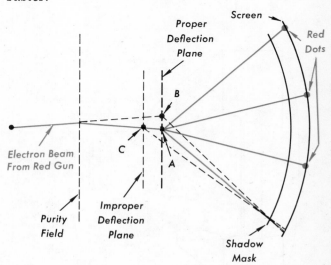

Fig. 9-17. Paths of the red beam as a result of proper and improper adjustment of the purity magnet and the deflection yoke.

If the beam enters the deflection plane at any other point, color impurity will result. Such a condition would occur if the red beam were to enter the deflection plane at point B in Fig. 9-17. Note that the dotted line from point B to the screen enters the shadow mask at a different angle than the line from point A to the screen. This condition can be corrected by the use of the purity magnet. The field of the magnet can be aligned so that the beam will enter the deflection plane at point A. The red beam is shown to be so aligned in the drawing.

The Deflection Yoke

The deflection yoke used with a three-beam color picture tube is shown in Fig. 9-18. Note that this yoke is very similar to those used in black-and-white television receivers. However, the core area is somewhat larger to allow sufficient space for the neck of the color picture tube. In addition, the protuding ends of the coils are sharply flared away from the core. This is done to prevent magnetic fields around these sections from interfering with beam focus or convergence action.

The position of the deflection yoke is a factor in obtaining color purity. This is illustrated in Fig. 9-17. Assume that the position of the yoke along

the neck of the tube is such that the red beam is deflected at point C. The dotted line from point C to the screen indicates the angle at which the beam would pass through the shadow mask. Since this angle is incorrect, color impurity would result. An improperly positioned deflection yoke will cause greater color impurity around the edges of the screen, since the error in the angle is greatest at those parts of the screen. The yoke must be moved forward or backward on the picture-tube neck until a position is found that provides the best purity at the outer edges of the picture area.

Beam-Positioning Magnets

The photograph in Fig. 9-19 shows the physical appearance of the beam-positioning magnets used in one make of color receiver. In the photograph, the magnets are spaced 120 degrees apart to correspond with the positions of the electron guns. These magnets are placed in a plane that is approximately perpendicular to the central axis of the picture tube and that intersects the three sets of pole pieces at the end of the gun assembly. The distance from any one of the magnets to the neck of the tube may be altered by sliding the shaft through the mounting.

The drawing in Fig. 9-20 shows the three fields that are produced by the positioning magnets. The strength of the field produced between each pair of pole pieces is dependent on the setting of the respective magnet. The force exerted upon each beam is perpendicular to the direction of the magnetic field between the pole pieces associated with

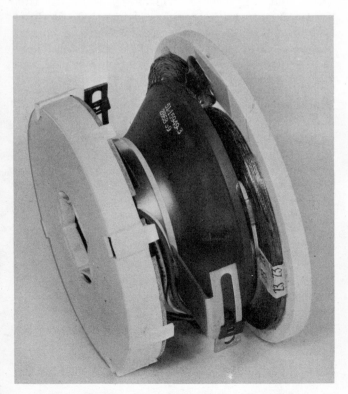

Fig. 9-18. Deflection yoke used with a three-beam color picture tube.

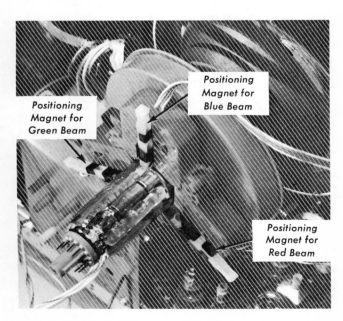

Fig. 9-19. Beam-positioning magnets.

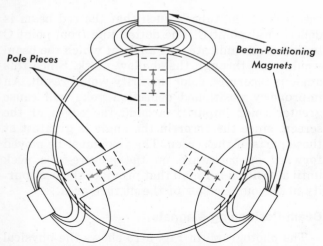

Fig. 9-20. Possible directions of the force exerted on each beam by the field of its associated positioning magnet.

that beam, as indicated by the solid arrows in Fig. 9-20. The amount of deflection on each beam is a function of the strength of the magnetic field between its associated pole pieces.

Fig. 9-21 shows another beam-positioning magnet that is cylindrical in shape and is mounted in a groove between the two sections of the ferrite core. The rod magnet is polarized, as indicated in the drawing, and produces a static field within its associated electromagnet. The magnet may be rotated so that a static field of either polarity and of the desired strength will be produced. Another variation of this principle is shown in Fig. 9-22 in which a rotatable disc performs the same function as the rod in Fig. 9-21.

Let us assume for the moment that the beams strike the shadow mask of a particular tube in the

Fig. 9-22. Beam-positioning magnets using rotatable discs.

pattern shown in Fig. 9-23. The dots represent the centers of the beams. The arrows indicate the directions that the beams can be moved by the fields produced by the beam-positioning magnets. Note that any two of the three beams in this arrangement can be converged but that all three beams cannot be converged; therefore, a fourth magnetic field is necessary. This additional field is provided by a device called the blue-lateral correction magnet.

Blue-Lateral Correction Magnet

The blue-lateral correction magnet is used to provide a field that will move the blue beam in a

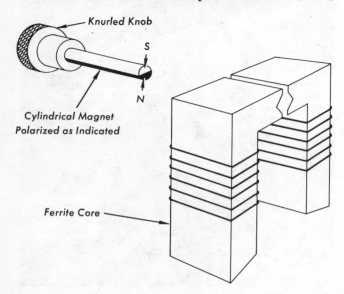

Fig. 9-21. Beam-positioning magnet using a rotatable rod.

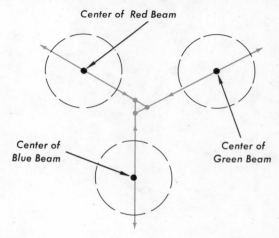

Fig. 9-23. Beams on the shadow mask and possible directions they can be moved by means of the positioning magnets.

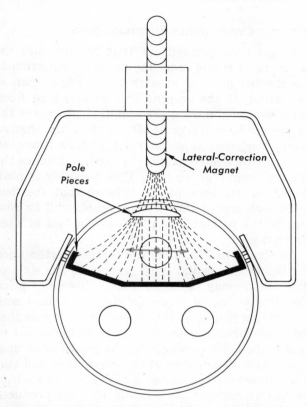

Fig. 9-24. The field of the blue-lateral correction magnet deflects the blue beam in a horizontal direction.

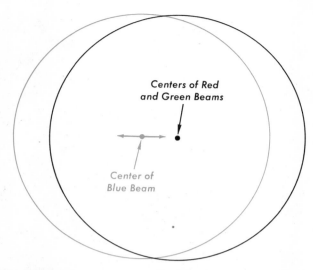

Fig. 9-25. Possible directions that the blue beam can be moved by means of the blue-lateral correction magnet.

horizontal direction. The blue-lateral magnet used with a 70-degree round color picture tube is shown in Fig. 9-24. This device is placed on the neck of the picture tube so that its field will intersect the special pole pieces mounted on the focus element of the blue gun. The magnetic field produced between these pole pieces and the direction of the beam movement caused by this field can be seen in Fig. 9-24. Now, we will see how the blue-lateral correction magnet aids in obtaining convergence of all these beams.

Consider the conditions shown in Fig. 9-23 in which the blue beam cannot be made to converge at the same point at which the red and green beams converge. Fig. 9-25 shows the points at which the beams can be made to strike the shadow mask by means of the positioning magnets. The blue arrows show the directions in which the blue beam can be moved by the field produced by the blue-lateral correction magnet. Note that the movement of the blue beam in the lateral direction will cause all three beams to be converged.

Two different types of blue-lateral correction magnets used with 70-degree round tubes are shown in Fig. 9-26. Both of these components produce the same effect. The magnet in the device la-

beled A in Fig. 9-26 is a rod that can be revolved 360 degrees. Therefore, a means is provided for varying the strength and reversing the direction of the magnetic field. The magnet used in the device labeled B can be moved in or out of its holder so that the distance between the magnet and the neck of the tube can be varied as needed. If a reverse field is required, the magnet can be reversed in the assembly.

The blue-lateral correction device used with a 90-degree, rectangular color picture tube is shown in Fig. 9-27. As mentioned previously, the blue gun used in the 90-degree color picture tube does not have the special pole pieces associated with the blue-lateral magnet. Fig. 9-28 shows the operation of the blue-lateral correction magnet used with a 90-degree color picture tube. The four magnets used with this device are positioned so that they produce a narrow magnetic field through the blue

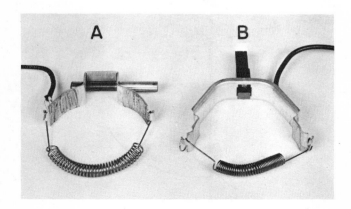

Fig. 9-26. Blue-lateral correction devices used with 70° round color picture tubes.

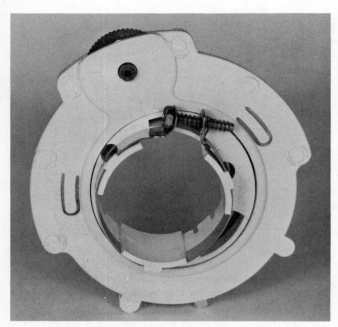

Fig. 9-27. Blue-lateral correction device used with 90°
rectangular color picture tube.

Dynamic-Convergence Electromagnets

It has been pointed out that the distance the beams must travel from the plane of deflection to the shadow mask is greater at the edges than at the center of the screen. The convergence force must be varied during the scanning process if the beams are to converge at all points on the shadow mask. For most types of color picture tubes, an assembly of electromagnets is used to produce the dynamic convergence field. This assembly is positioned on the neck of the tube so that the field from each one of the electromagnets will be coupled to the pair of pole pieces at the end of each electron gun.

As shown in Fig. 9-29, the forces exerted upon the beams are radial with respect to the axis of the tube. It may be recalled that this description also applies to the forces produced by the beam-positioning magnets; therefore, it can be said that the fields produced by the electromagnets aid or oppose the fields produced by the positioning magnets. Dynamic voltages at the horizontal- and vertical-scanning frequencies are applied to each of the electromagnets. As a result, the field produced by a positioning magnet and the field produced by the associated electromagnet will aid each other when current through the electromagnet flows in one direction and will oppose each other when the current flow is reversed. This being the case, the beams can be independently directed to converge on the shadow mask at all times during the scanning process.

Again referring to Fig. 9-29, the reader will note that the electromagnets have horseshoe-

gun. Note that the field passing through the blue gun is narrow enough that it does not affect the red or green guns.

When the adjustment wheel is turned, the magnets move farther apart or closer together, depending in which direction the wheel is turned. The magnetic field becomes stronger as the magnets move closer together and it becomes weaker as the magnets are moved farther apart. The field may be reversed by rotating the assembly 180 degrees.

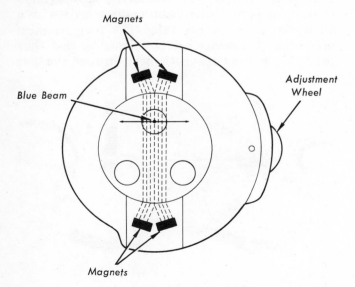

Fig. 9-28. Field produced by blue-lateral correction
device shown in Fig. 9-27.

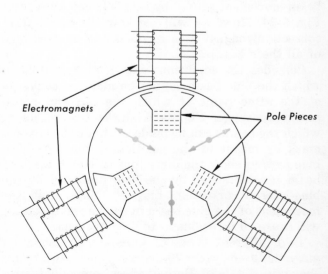

Fig. 9-29. Possible directions of the forces exerted on each
beam by the field of its associated electromagnet.

shaped cores. These cores are constructed so that they can be aligned with the pole pieces in the tube. There are two separate windings on each core, and the dynamic convergence voltages are applied to these windings. The nature of these voltages and their derivations will be discussed later in this chapter.

Fig. 9-30 shows the electromagnets and the method of mounting them on a 25-inch rectangular tube. This mounting arrangement is typical in re-

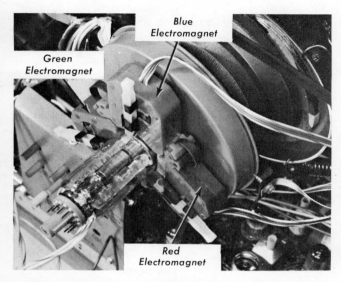

Fig. 9-30. Electromagnets used for dynamic convergence.

ceivers that employ this type of picture tube. The physical shape of many auxiliary devices will vary, but the basic principles remain the same.

SINGLE-GUN PICTURE TUBE

One of the newest features in the color television industry has been the development of the Trinitron—a one-gun picture tube. This tube is being used by Sony in their solid-state color television. Whereas the conventional three-gun color tube has the color phosphors arranged in dot triads, the Trinitron has the color phosphors arranged in thin vertical stripes. An aperture grille, which has one slot for each group of three primary-color stripes, is positioned behind the phosphor stripes (Fig. 9-31). It can be seen that the diameter of a single beam spans about two slots of the aperture grille. Thus, the electrons strike the centers of two phosphor stripes.

In Fig. 9-32 is a simplified drawing of one beam and one primary-color phosphor. The beam illustrated is the green beam which is emitted from the center of the gun.

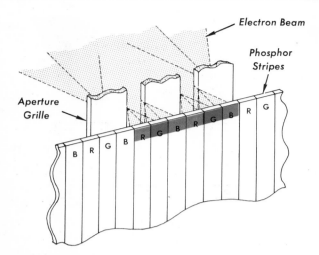

Fig. 9-31. Position of the aperture grill in relation to the phosphor stripes.

The beams intended to strike the red and blue phosphors approach the aperture grille from either side of the green beam. Fig. 9-33A shows the electron path of the red beam which is emitted from the gun to the right of the green beam. The beam is directed through the aperture grille to strike two narrow spots at the center of the red phosphors. In a similar manner, the electrons intended for the blue phosphors approach the aperture grille from the left of the green beam, as shown in Fig. 9-33B.

Electron-Gun Assembly

A comparison between the gun assembly used in most color picture tubes and the one used in the Trinitron tube is shown in Fig. 9-34. The beams from the Trinitron gun assembly are "in line." There are certain advantages to this. It greatly simplifies dynamic convergence. Also, because of a common lens assembly, the gun is smaller and can fit into a tube with a much smaller neck diameter

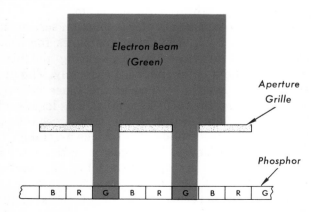

Fig. 9-32. Action of the center (green) beam on the green phosphors.

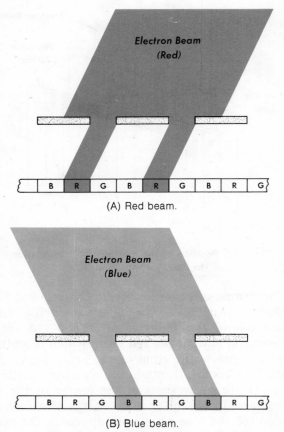

(A) Red beam.

(B) Blue beam.

Fig. 9-33. Action of the red and blue electron beams.

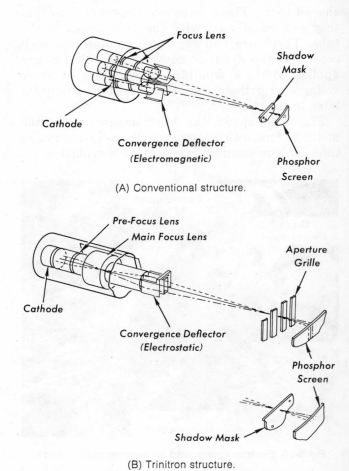

(A) Conventional structure.

(B) Trinitron structure.

Fig. 9-34. Comparison between conventional and Trinitron electron-gun structures.

than that of the conventional color tube. Because of the small gap between the poles of the deflection coils, less deflection current is required.

Fig. 9-35A shows a simplified sectional drawing of the gun assembly. The gun has three cathodes with individual heaters. The three cathodes are enclosed in a single control-grid cup (G1), with one aperture for each beam. Signal voltage applied between the individual cathodes and the common control grid (G1) controls beam intensity, and thus the brightness of each beam.

Next, the beams encounter a common screen cup (G2). The G2 cup has three apertures for beam passage and provides an accelerating potential. The beams next enter the focus assembly. The first electrostatic lens is formed in the space between the G2 cup and the focus assembly. This lens bends the red and blue beams to cross the green (central) beam in the middle of the focus assembly. To simplify the action of the electron optics system, Fig. 9-35B shows an optical model of the electrostatic lens.

Beam focus is accomplished by a focus assembly which is very similar to that of the unipotential guns used in black-and-white picture tubes. The

assembly acts like a large-diameter lens whose focal length is selected to make each beam focus at a fine spot on the phosphor screen. Refer to Fig. 9-35B. A great advantage of the Trinitron picture tube is that the first lens diverts all three beams through the center of the large focus lens. In effect, it gives the equivalent of a large depth of field. As a result, the beam stays in focus regardless of variations in distance between the gun and screen. In this manner, uniform focus can be achieved at both the edges and the center of the screen.

The three beams leave the focus assembly in slightly divergent paths. They now pass between a set of four static convergence plates, as shown in Fig. 9-35A. Because there is no difference in potential between the two center plates, the green beam passes through unaffected. However, a static voltage is applied between the outer and inner plates. The action of the fields existing between the plates is to cause the red and blue beams to converge at the aperture grille.

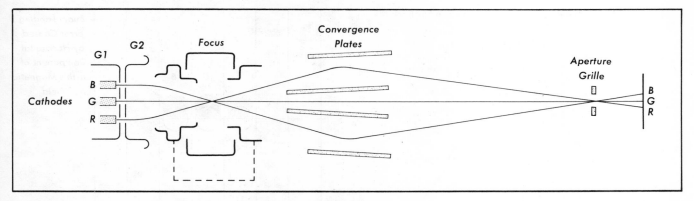

(A) Top view.

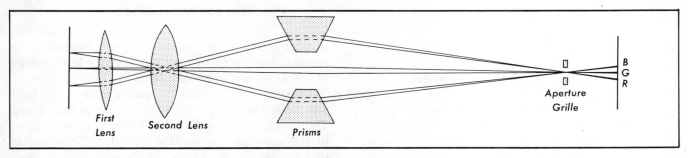

(B) Optical model of lens system.

Fig. 9-35. Simplified sectional drawing of Trinitron gun assembly.

Beam-Landing Controls

For the electron beams to land correctly, they must approach the aperture grille at the correct angles. Refer to Fig. 9-35A. To accomplish this, the three beams must originate from specific locations with respect to the aperture grille-phosphor plate assembly.

To establish proper beam landing at the sides of the screen, the point along the axis at which deflection takes place must be placed accurately. This point is termed the deflection center. It is located at the approximate center of the deflection yoke. This point can be moved along the tube's axis by positioning the deflection yoke along the neck of the tube. Fig. 9-36 illustrates correct and incorrect beam landings. Incorrect landing is shown by the dotted lines.

As we can now see, two adjustments are needed to make the electrons land on their designated phosphors. One adjustment, accomplished by the purity magnet assembly, affects the beam landing at the center of the screen (Fig. 9-37). A control on the assembly moves the three beams from left to right. The second adjustment, which corrects beam landing at the sides of the screen, is accomplished by yoke positioning.

As seen in Fig. 9-38, vertical displacement of the beams will not result in the beams hitting the wrong phosphors. This simplifies beam-landing adjustments in that only horizontal correction is needed.

Neck-Twist Coil

Although the neck-twist coil affects beam landing, it is actually adjusted for best convergence. The adjustment is made to correct convergence on the horizontal lines of a crosshatch pattern. The action of this coil is to align the outer beams around the center green beam (Fig. 9-39). The coil current and polarity are adjusted to shift the three beams into a horizontal plane relative to the gun assembly (Fig. 9-40).

Dynamic Convergence

Because the distance between the deflection centers and the aperature grille increases as the beams move away from the center of the screen, the beams converge short of the aperture grille at the edges of the tube. In conventional color picture tubes, correction is made by applying dynamic convergence correction, which is synchronized with the scanning signals.

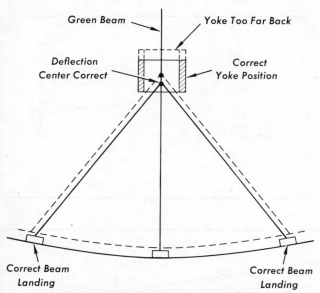

Fig. 9-36. Proper and improper beam landings.

In the Trinitron system, dynamic convergence correction is simplified because of the electron-beam arrangement. Vertical correction is symmetrical because all beams are in the same horizontal plane. To effect horizontal convergence, a parabolic waveform, synchronized with the horizontal scan, is developed in the horizontal-output stage. This waveform is added to the dc voltage applied to the outer deflection plates. In order to divert the outer beams, the outer plates are normally at a negative potential with respect to the inner plates. As the beams sweep toward the sides of the screen, the dynamic convergence voltage reduces the potential difference between the inner and outer plates, thus "straightening out" the three beams and making them converge at a

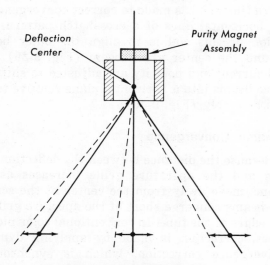

Fig. 9-37. Action of purity magnet.

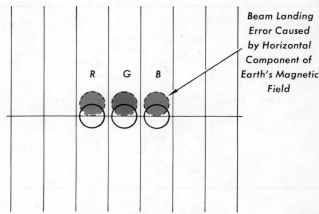

Fig. 9-38. Beam-landing error.

greater distance from the deflection center. In essence, the beams spread out slightly to converge at the sides of the screen (Fig. 9-41).

Vertical convergence is accomplished without any added circuitry because of the in-line guns. The vertical coils are toroidal-wound as compared to the conventional saddle winding. This results in a barrel-shaped vertical-deflection field as seen in Fig. 9-42. In the center of the screen where the flux lines are straight and the beam deflection is in the vertical direction only, there is no convergence correction.

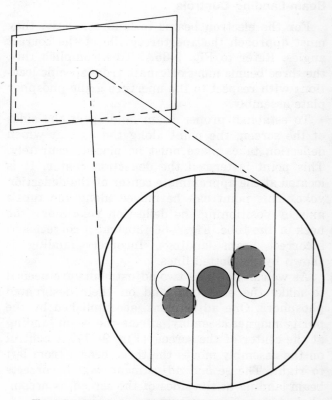

Fig. 9-39. Effect of neck-twist coil on beam landings.

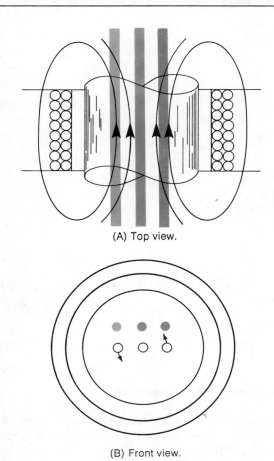

(A) Top view.

(B) Front view.

Fig. 9-40. Neck-twist coil.

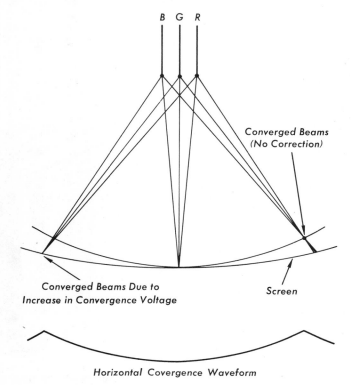

Fig. 9-41. Horizontal dynamic convergence.

At the top of the picture the flux lines are bent and have a vertical component as well as a horizontal component, and the flux line is rising for the left beam and falling for the right beam. This vertical magnetic-field component contributes some horizontal deflection to the outer beams. This deflection tends to divert the beams, causing them to converge at a greater distance from the deflection centers. This same effect also takes place at the bottom of the picture. In this manner, vertical convergence can be attained without additional controls or circuits.

OTHER IN-LINE PICTURE TUBES

Several manufacturers now have picture tubes with in-line guns and striped screens. One such tube, produced by RCA, is shown in Fig. 9-43. This color picture tube is similar to the Trinitron inasmuch as its electron-gun assembly employs three separate cathodes with single control-grid (G1), screen-grid (G2), focus (G3), and high-voltage-anode (G4) elements. Each of these elements has three apertures, one for each beam. A simplified drawing of the electron-gun structure is shown in Fig. 9-44.

As shown in Fig. 9-43, the shadow mask used with the RCA tube is somewhat different from the shadow mask used with the Sony Trinitron. A series of small cross-ties are placed across the slots

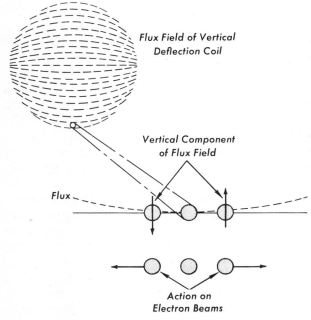

Fig. 9-42. Vertical dynamic convergence.

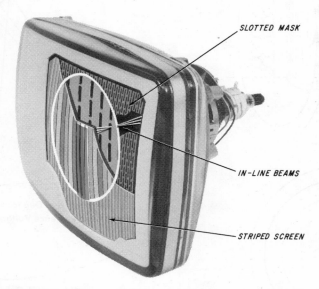

SLOTTED MASK

IN-LINE BEAMS

STRIPED SCREEN

Fig. 9-43. RCA color picture tube employing in-line gun and striped phosphor screen.

to give the mask more rigidity. Also, the placement of the phosphor stripes is different in relation to the electron beams. In the RCA tube, the center beam strikes the red stripe, while the center beam of the Trinitron tube strikes the green stripe. Another notable difference is the method of beam control.

A unique feature of the RCA tube is the precision-wound-static-toroid deflection yoke. The construction of this yoke is quite different from conventional deflection yokes. Because each turn of wire is accurately positioned on the yoke core, the field developed by this yoke will produce a linear raster that is converged over the entire screen. In addition, the raster is essentially free of pincushioning. The small neck diameter of this picture tube makes it possible to provide full deflection with smaller yoke currents. The low impedance of this yoke is readily matched to sold-state sweep circuits.

The precision design of the electron gun used in the RCA in-line tube ensures that the three beams are accurately aligned with respect to each other. However, during the assembly of the picture tube, it is possible for gun-assembly alignment errors to occur with respect to the aperture mask and the striped phosphor screen. Therefore, it is necessary to provide some means for obtaining good purity and static convergence. This is achieved by exact

positioning of the deflection yoke and correct adjustment of the convergence and purity assembly mounted on the neck of the tube. The deflection yoke and the convergence and purity assembly are shown in Fig. 9-45.

Proper positioning of the deflection yoke and correct adjustment of the convergence and purity assembly are included in the manufacturing process. After the deflection yoke has been properly positioned, it is cemented in place. Since it cannot be removed, the deflection yoke is considered an integral part of the picture tube. The convergence and purity assembly is then adjusted and locked into place. This component normally does not require readjustment unless the locking ring comes loose.

Several other manufacturers produce in-line picture tubes, although their operation may be somewhat different. However, due to simpler convergence adjustments and reduced convergence circuitry, it is expected that the in-line color picture tube will be widely used in the near future.

DYNAMIC CONVERGENCE CIRCUITS

Before we discuss the convergence circuits, let's take a look at the reasons behind all of the special circuitry and setup procedures. Fig. 9-46 is a description of the convergence error caused by the

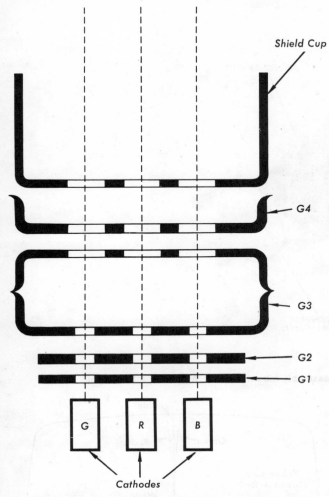

Fig. 9-44. Electron-gun assembly used in the RCA in-line picture tube.

and the horizontal line crossing at the center of the picture tube face) are of importance. The reason for observing this rule is that any change in the vertical lines or row of dots at the center of the screen is the result of vertical convergence signals; the horizontal convergence signals at this point are at zero. Also, any change in the horizontal lines or dots across the center of the screen is the result of horizontal convergence signals, because at this time the vertical convergence signals are at zero. Beam convergence in any of the four quadrants is controlled by a combination of both vertical and horizontal convergence signals.

Currents in the Convergence Coils

The drawing in Fig. 9-48 shows the points of convergence for the three beams and also indicates the amount of correction needed at the limits of the sweep. The three beams should be converged at the center of the screen. Convergence at the center is accomplished by adjusting the static-convergence magnets composed of the three beam-positioning magnets and the blue-lateral correction magnet.

The waveshape of the current required to correct the convergence of the beams must be parabolic for the horizontal and vertical sweep as shown in Fig. 9-49A. Provision must be made for increasing or decreasing the amplitude of the current as shown in Fig. 9-49B. It is also necessary to tilt the waveform to the right or left as shown in Fig. 9-49C.

The basic method of producing the necessary waveshapes is to supply a parabolic current to the coil and then add a sawtooth wave to provide the desired tilt. The result of adding a sawtooth wave to a parabolic wave to provide tilt to the left or right is shown in Fig. 9-50. The amplitude of the sawtooth wave determines the degree of tilt, and the direction of the slope determines whether the tilt is to the left or to the right.

The vertical convergence signals and the horizontal convergence signals are usually applied to separate coils on the convergence electromagnet. One of the dynamic convergence electromagnets is shown in Fig. 9-51. The large windings are the vertical convergence coils, and the small windings are the horizontal convergence coils.

Circuit Action

The convergence circuitry used in a color receiver is shown in Fig. 9-52. Notice that the horizontal and the vertical circuitry are separate. The signals are applied to separate windings on the

basic construction of the color picture tube. As previously pointed out, the point of convergence for all three beams describes an arc about the deflection center. Proper convergence of all three beams can be obtained at the center of the picture tube, but as the beams are deflected away from the center, the point of convergence moves back from the shadow mask. The effect on the screen of a vertical and horizontal line with no convergence signals applied is shown in Fig. 9-47.

Notice that there are three *vertical* lines and only two *horizontal* lines. An examination of Fig. 9-46 will show that a vertical deflection of the three beams will produce three parallel lines on the screen, whereas a horizontal deflection will cause the red and green beams to traverse the same path across the screen and produce a yellow line.

When the effect of the convergence signal is being observed, only two areas (the vertical line

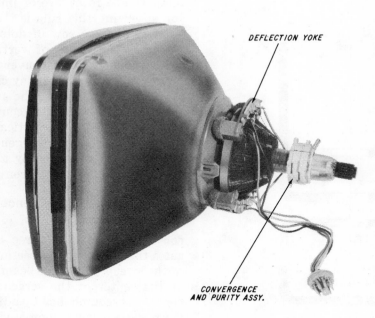

Fig. 9-45. Deflection yoke and convergence and purity assembly mounted on RCA in-line picture tube.

electromagnet, and since the core is common to both vertical and horizontal coils, the fields of both are combined.

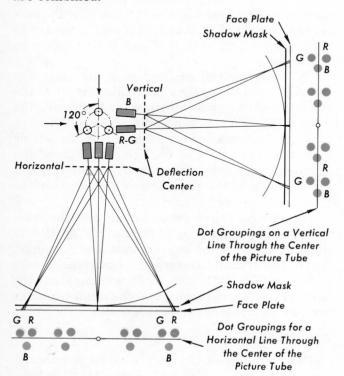

Fig. 9-46. Illustration of the convergence error caused by the position of the guns and the direction of sweep.

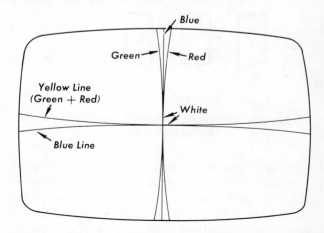

Fig. 9-47. Effect caused by a lack of dynamic convergence on a vertical line and a horizontal line through the center of the color picture tube.

Vertical Convergence—The operation of the convergence circuit in Fig. 9-52 can best be described by starting with a basic circuit, such as

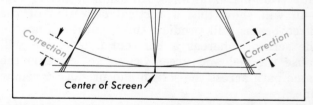

Fig. 9-48. Convergence correction needed at the limits of vertical and horizontal sweep.

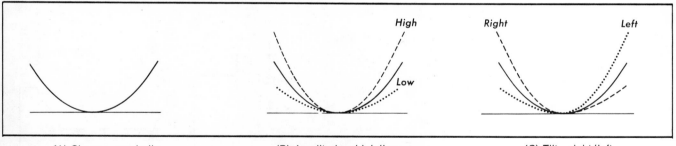

(A) Shape—parabolic.   (B) Amplitude—high/low.   (C) Tilt—right/left.

Fig. 9-49. Waveforms of current supplied to the convergence coils.

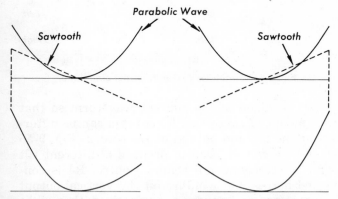

Fig. 9-50. Result of adding parabolic and sawtooth currents.

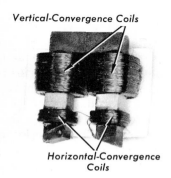

Fig. 9-51. Photo of a dynamic convergence electromagnet.

Fig. 9-52. Convergence circuit used in a color receiver.

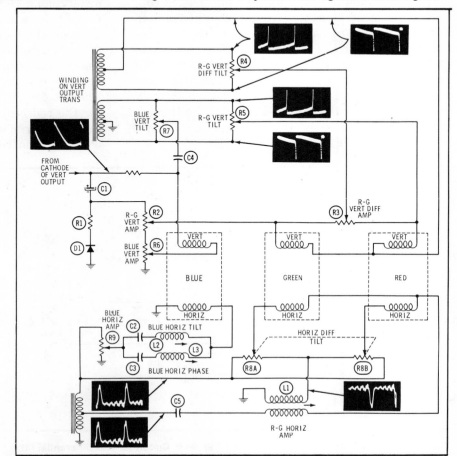

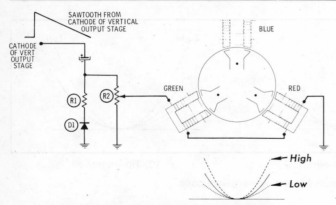

Fig. 9-53. Simplified circuit for R-G vertical amplitude.

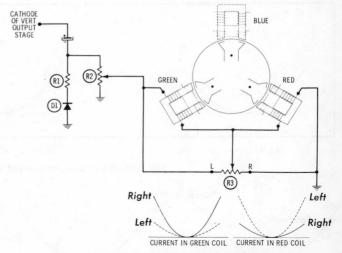

Fig. 9-54. Circuit for R-G differential amplitude.

the one shown in Fig. 9-53. The parabolic current for the convergence coil is derived from a sawtooth wave obtained from the cathode of the vertical-output stage. Diode D1 and series resistor R1 serve to modify the sawtooth so that a parabolic current will flow in the coil. Control R2 adjusts the amplitude of the current through the red- and green-convergence coils. The direction of the current in both coils is the same. In order to get a different amplitude in each coil, a differential control is added, as shown in Fig. 9-54. Control R3 adjusts the amount of current flowing in each coil. Moving the control arm of R3 to the right tends to short out the red coil and at the same time increase the resistance across the green coil. The result is to increase the current through the green coil and decrease the current through the red coil. The amount of current through R2 remains relatively constant. Moving the control arm of R3 to the left reverses the balance of current in the red- and green-convergence coils.

Tilt is added to the current waveform so that the amount of correction for a beam can be different at the top and bottom of the screen. Fig. 9-55 shows the circuit used to produce a different tilt for the green and red beams. Control R4 is connected across one winding on the vertical-output transformer and causes a sawtooth, with a right or left slope, to be applied to the coils. The current from this winding flows in opposite directions through the coils as shown by the arrows in Fig. 9-55. When the control arm is in the positions shown, the currents through the convergence coil are as shown by the broken lines on the waveform. The tilt for the green beam is to the left, and tilt for the red beam is to the right. The solid line on the waveform indicates that no tilt is introduced when the control is at the center position.

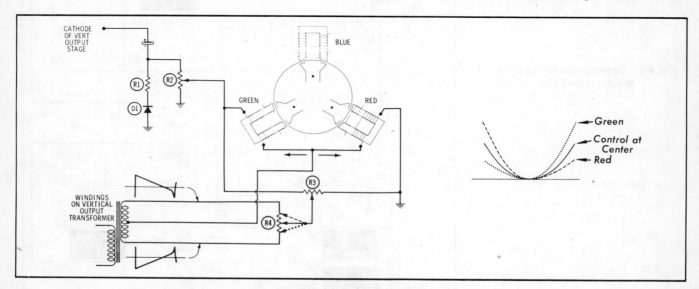

Fig. 9-55. Circuit for R-G differential tilt.

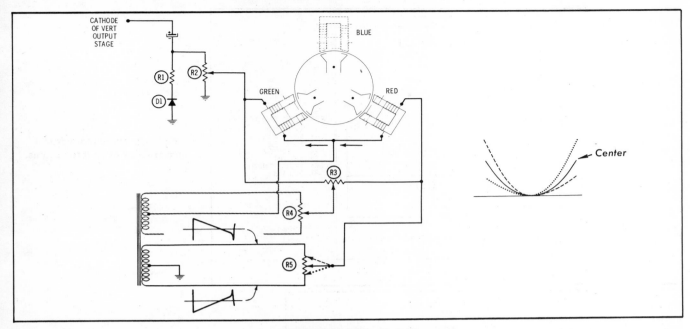

Fig. 9-56. Circuit for R-G tilt.

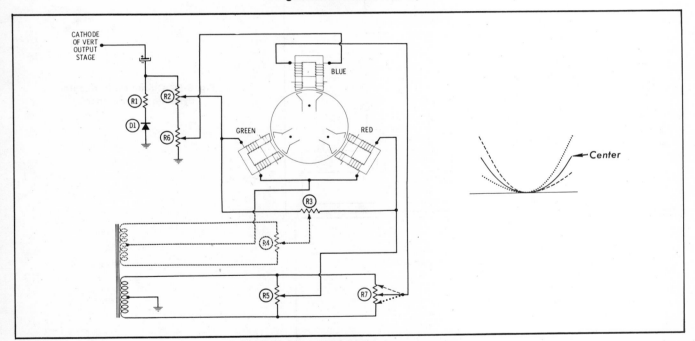

Fig. 9-57. Circuit for blue vertical convergence.

Another tilt circuit for the vertical red and green convergence is shown in Fig. 9-56. The direction of tilt for red and green lines and the direction of current through the coils at any instant is the same. The positions of the dotted, dashed, and solid arrows on control R5 correspond to the dotted, dashed, and solid curves that indicate the current through the convergence coils.

The five controls just discussed affect only the

red and green beams. The principles employed in these five examples are used throughout the convergence circuits with a few variations.

The additional circuitry for the blue-vertical convergence is shown in Fig. 9-57. Control R6 determines the amplitude of the parabolic current applied to one end of the vertical-convergence coils for the blue gun. The other end of the winding is connected to the movable arm of blue-vertical tilt

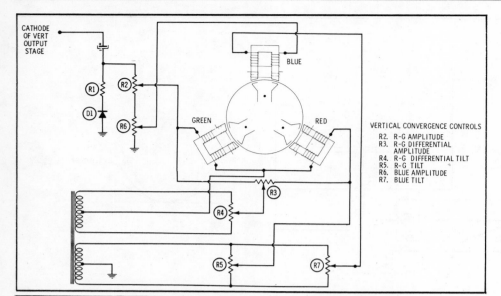

Fig. 9-58. Circuit for vertical convergence of all three beams.

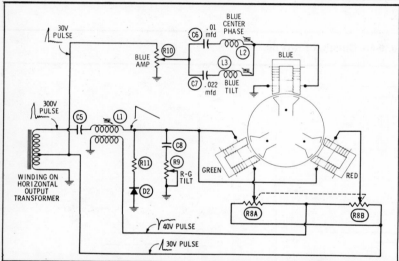

Fig. 9-59. Circuit for horizontal dynamic convergence.

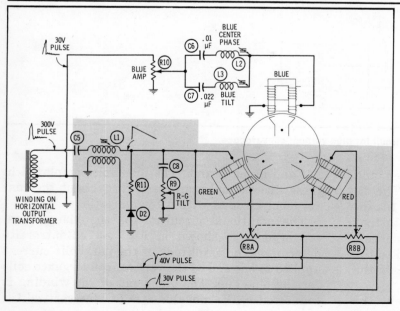

Fig. 9-60. Circuit for horizontal convergence of green and red beams.

control R7. This control impresses a sawtooth current onto the parabolic current, causing the vertical line to tilt to the left or right.

A schematic of the entire vertical convergence circuit is shown in Fig. 9-58. The controls listed in this figure are named according to their function. In many receivers, the controls are labeled according to the effect they have upon a cross-hatch or dot pattern to make it easier to follow a certain procedure.

Horizontal Convergence—The controls for the horizontal convergence circuits are somewhat different from the vertical convergence controls due to the higher frequency and the available waveshapes used in horizontal convergence.

The circuit for horizontal dynamic convergence is shown in Fig. 9-59. This circuitry is less involved than that shown for the vertical convergence because the red and green guns occupy similar positions relative to the horizontal sweep. The relative positions of the three guns are shown in Fig. 9-46. Also, the paths traversed by the beams across the face of the tube are shown in Fig. 9-47. Notice that the red and green beams follow the same path, and therefore it is possible to have fewer controls and still obtain proper horizontal convergence.

In the circuit of Fig. 9-60, the current for the red and green coils is a 300-volt pulse supplied from a winding on the horizontal-output transformer. Tuned inductor L1 shapes this pulse into a sawtooth waveform which produces a parabolic current in the green and red convergence coils. Controls R8A and R8B are ganged potentiometers,

connected so as to return the coils to ground through either of two windings. One winding supplies a positive pulse and the other a negative pulse. With the controls at approximate center, the two pulses cancel and have no effect on the current in either coil. When the control is moved, one coil receives the positive pulse and the other coil receives the negative pulse. This causes a different sawtooth current in each coil and produces a different tilt for each beam. When the control is moved in the opposite direction, the tilt current is reversed in both coils.

The R-G tilt control, R9, increases the resistive loading on the red and green coils, causing the parabolic current to take on some of the characteristics of the applied sawtooth wave. This has the effect of tilting the parabolic current the same direction in both the red and green convergence coils.

The circuit for blue horizontal convergence is shown in Fig. 9-61. A 30-volt pulse is supplied from the winding on the horizontal-output transformer. Coil L3 and capacitor C7 are tuned to resonance at the horizontal scan rate of 15,750 Hz. Coil L2 and capacitor C6 are tuned to resonance at the second harmonic of the horizontal scan rate, or 31,500 Hz. The combination of these tuned circuits provides a parabolic current to drive the blue-convergence electromagnet. Blue tilt is produced by tuning L3, which changes the phase relationship between the two tuned circuits. The amplitude of the current through the blue coil is controlled by potentiometer R10, and since the same amplitude signal is supplied to both tuned circuits, it does

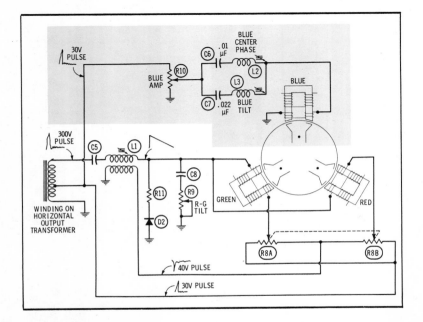

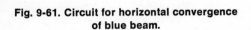

Fig. 9-61. Circuit for horizontal convergence of blue beam.

not disturb the preset phase relationship between the two circuits.

The convergence circuits in different makes and models will vary considerably, but in receivers that employ the conventional three-gun color tube, the circuitry and the basic theory will remain much the same. Each manufacturer has a particular alignment sequence that has been developed for adjusting the color-convergence controls. It is recommended that the procedure outlined by the manufacturer be used in each case. In most instances, the tuning of convergence controls out of sequence will result in misalignment that can be corrected only by a complete resetting of the controls according to the manufacturer's procedure.

QUESTIONS

1. Describe the construction of the color picture-tube screen.
2. What is the position of the blue gun when the blue phosphor dot is below the red and green dots?
3. Which of the following terms refers to the process in which the beam of each gun is adjusted to fall on its respective color dot?
   (a) Focus
   (b) Deflection
   (c) Convergence
   (d) Phasing
4. Beam-positioning magnets affect all three beams at the same time. True or false?
5. Name two basic types of convergence adjustments used with the three-gun color tube.
6. What is the purpose of the cone shield?
7. What adjustment is made to ensure that the beams will enter the deflection field properly?
8. What additional adjustment device is provided for the blue gun only?
9. What type of convergence is supplied by the three electromagnets around the tube neck?
10. What is the shape of the horizontal-convergence voltage? Choose one.
    (a) Sawtooth
    (b) Sinusoidal
    (c) Square wave
    (d) Parabolic
11. From which circuits are the dynamic convergence signals obtained?

Section III

Servicing the Color Receiver

Chapter 10

Setup Procedure

The ultimate goal of the setup procedure is to ensure that the picture tube will be capable of producing the correct colors when a color signal is being received and that it will reproduce a black-and-white picture when a monochrome signal is being received. In order to achieve this goal, the beams must be controlled so that each one will strike the correct set of phosphor dots and so that the beams will cross correctly at the plane of the aperture mask. With each beam striking the correct set of phosphor dots, color purity is achieved. When the three beams cross at the plane of the aperture mask, the beams are properly converged. After proper purity and convergence are achieved and the intensities of the beams are in balance to produce the proper gray scale, the tube should produce good color pictures and good monochrome pictures.

PRELIMINARY ADJUSTMENTS

Before proceeding with the adjustments that are particularly associated with the color picture tube, it is advisable to check the operation of several other circuits. A test signal from a pattern generator is very useful for this purpose.

The operation of the high-voltage circuit must be checked in the preliminary procedure. Since the color picture tube must operate under extremely exacting conditions, the high-voltage supply must produce a constant voltage output. A voltage-regulator circuit is employed for this purpose. This circuit, in many receivers, includes a control that should be adjusted so that the high-voltage output has the value recommended by the manufacturer.

If the high voltage in some color receivers is allowed to go beyond the value recommended by the manufacturer, the result can be excessive ra-diation or even a damaged picture tube. The high-voltage adjustment control has been eliminated in some of these receivers and the high-voltage limits are determined by circuit design. When a high-voltage control is provided, it is important to adjust the control exactly as specified by the manufacturer.

When the brightness is advanced beyond its normal range, the high-voltage regulator circuit may lose control, causing the ultor voltage to decrease considerably. This abnormal brightness might cause the control grids of the picture tube to draw current. Many receivers have a brightness-range or brightness-limit control to limit the range of the brightness. If such a control is provided, it should never be allowed to remain at a setting that will provide abnormally high brightness. Again, this control should be adjusted according to the manufacturer's recommendations.

After the high voltage has been properly adjusted, the conventional adjustments pertaining to picture size, linearity, vertical and horizontal centering, and focus should be made. These adjustments should precede those pertaining to purity and convergence, because later changes in the sweep or high-voltage circuits may necessitate a repetition of the purity and convergence adjustments.

In the preliminary setup procedure, adjustments are made in approximately the following order:

1. Tune in a signal and set the color control at minimum.
2. Check the lock-in range of the horizontal-oscillator circuit, and make any necessary adjustments.
3. Adjust the agc voltage for maximum picture contrast without distortion.

4. Turn the contrast and brightness controls to the recommended settings.

5. Use a suitable high-voltage probe, and measure the ultor voltage to see if it compares with the value specified by the manufacturer.

6. Rotate the brightness control through the range that is normally used, and note the variation in ultor voltage.

7. If the variation is excessive, make the recommended high-voltage adjustments.

8. Adjust the horizontal linearity and centering, if provided.

9. Adjust the vertical size, linearity, and centering.

10. Adjust the focus control.

DEMAGNETIZING THE COLOR PICTURE TUBE

The three-gun color picture tube is sensitive to variations in magnetic fields, and the earth's magnetic field is of particular importance. Although the earth's magnetic field is relatively weak, it is this field we are primarily interested in. When a receiver is moved from one location to another, or the position the receiver faces is changed, the receiver's relationship to the earth's magnetic field changes. This change can affect the convergence of the dots on the picture-tube screen.

Most receivers have automatic degaussing circuits that demagnetize the receiver each time it is turned on. Whether the receiver has automatic degaussing or not, when the convergence adjustments are being made, it should be regular practice to demagnetize the color picture tube using a degaussing coil such as the one shown in Fig. 10-1. A degaussing coil can be purchased or one can be wound, using 425 turns of No. 20 wire made into a ring approximately 12 inches in diameter. The power cord should be about nine feet in length.

Fig. 10-1. Degaussing coil.

The receiver should be placed in the position that it will occupy when it is being viewed. The degaussing coil should be placed at a distance of about six to ten feet from the face of the receiver and then plugged into the wall outlet. Approach the receiver with the coil turned on and pass the coil, using a circular motion, over the face of the tube and the front and sides of the cabinet. Continue the circular motion and move back at least six feet from the receiver before unplugging the coil.

ADJUSTMENTS FOR STATIC CONVERGENCE

Prior to making any adjustments for color purity, the static convergence adjustments should be made. When this has been done, the three beams of the color picture tube will converge at the center of the shadow mask. Any misadjustment of a beam-positioning magnet may cause its associated beam to be displaced too far from the proper position. Under such a condition, it would be impossible to obtain color purity. For this reason, static convergence adjustments should precede the purity adjustments to make it possible for all three beams to produce pure rasters with the same setting of the purity adjustments.

The adjustments for obtaining static convergence are made while a white-dot pattern is being observed on the picture tube (Fig. 10-2). Three-color dot patterns are produced by the three beams as a result of the video signal applied to the picture tube. The dots in the patterns are red, green, and blue. These colors in the dots may not be pure toward the edges of the screen, since the purity

Fig. 10-2. Pattern produced on the picture-tube screen
by a white-dot generator.

adjustments will not have been made at this point in the setup procedure. Impure colors in the dots toward the edges of the screen are therefore disregarded when making the preliminary adjustments to obtain static convergence.

It is possible that the dot produced by one of the beams will not be clearly visible. This condition is likely to occur when the voltages applied to the elements of one of the picture-tube guns are considerably different from those applied to the other two guns. In order to compensate for this condition, the screen control for the gun producing the dim dot may be advanced until the associated color phosphor dots are more brightly illuminated.

The controls used to obtain dynamic convergence may be set for minimum correction before the static convergence adjustments are made. However, in most instances the controls will be very close to the proper position, and it is not necessary to adjust them for minimum correction.

The usual practice is to perform the convergence procedure, then return to the beginning and repeat the entire procedure. During the repeat adjustments, each control should require only a small touch-up, but the final results are generally excellent.

Four devices are used to obtain static convergence of the beams in the conventional color picture tube. These are the three beam-positioning magnets and the blue-lateral correction magnet. The blue-lateral correction magnet for a 70-degree tube should be positioned in line with the pole pieces mounted on the focus element of the blue gun, and the magnetic-convergence assembly should be positioned so that the cores of the electromagnets will be aligned with the pole pieces at the end of the gun assembly. As mentioned previously, the 90-degree rectangular picture tubes do not have these special pole pieces mounted on the blue gun. Consult the service data for correct positioning of the blue-lateral correction device used with a 90-degree rectangular picture tube.

The directions in which the beams may be moved by the four static convergence devices are shown by the arrows in Fig. 10-3. These directions also apply to any pattern produced by the beams; consequently, the patterns produced on the screen by the signal from the dot generator can be moved so that a reference dot in the center of one color pattern will coincide with the corresponding center dot of each of the other two patterns. The combined light emissions from the three color phosphors will produce white light; therefore, the three coinciding dots of light at the center of the

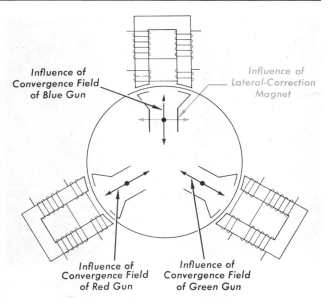

Fig. 10-3. Directions in which beams can be moved by convergence fields.

screen will appear as one white dot. When this has been accomplished, static convergence of the beams will have been achieved. Other dots near the center of the screen will appear white, but dots away from the center will not be converged and will have color fringing.

Fig. 10-3 shows that the directions in which the red and green beams may be moved by their associated convergence fields are such that the dots of light from these beams will coincide only when the associated beam-positioning magnets are at specific settings. The blue beam, on the other hand, can be moved both vertically and horizontally. For this reason, the red and green beams should be converged first. Then, through the use of the beam-positioning magnet which is associated with the blue gun and the blue-lateral correction magnet, the blue dots of light can be moved so that they will coincide with the red and green ones.

Since the combined light emissions from the red and green phosphor dots will give the appearance of yellow, the dots of light at the center of the screen will appear to be yellow when the positioning magnets for the red and green beams have been adjusted properly. The yellow dots shown in Figs. 10-4 and 10-5 indicate that these two magnets have been adjusted correctly. Fig. 10-4 shows, in addition, the blue dots displaced horizontally from a position of static convergence by a misadjustment of the blue-lateral correction magnet. Fig. 10-5 shows the blue dots displaced vertically by a misadjustment of the beam-positioning magnet associated with the blue gun.

Fig. 10-4. Blue dots are displaced horizontally by misadjustment of the blue-lateral magnet.

The order in which the adjustments for static convergence should be made is as follows:

1. Supply a white-dot signal to the receiver.
2. Adjust the contrast and brightness controls for good pattern reproduction on the screen.
3. Adjust the screen controls for the red, green, and blue guns until each of the three colored dots is clearly visible.
4. Set the dynamic convergence controls for minimum correction only if necessary.
5. Position the magnetic-convergence assembly and the blue-lateral correction magnet so that they will have the proper relationship to the pole pieces in the gun assembly. It may be necessary to consult the service data to

determine proper positioning for these devices.
6. Converge the red and green dots of light at the center of the screen by adjusting the beam-positioning magnets for the red and green guns.
7. Adjust the beam-positioning magnet for the blue gun, and adjust the blue-lateral correction magnet to converge the blue and yellow dots of light at the center of the screen.

During the purity and dynamic-convergence adjustment procedure, it may be necessary to repeat the static convergence adjustments. Also, in many cases of color receivers being out of convergence, the trouble can generally be remedied by simply adjusting the four static convergence magnets. These adjustments are of a mechanical nature, and the vibration and shock produced by moving the receiver will tend to change these adjustments more than others in the receiver.

The static convergence adjustments that have just been discussed are for conventional color picture tubes employing the triad dot arrangement. The static convergence adjustments for the in-line or single-gun color picture tubes are somewhat different from the procedure for conventional tubes. Also, these procedures vary considerably from one tube type to another. Therefore, it is not possible to give a general static convergence procedure that would apply to the various types of in-line picture tubes now in use. However, the convergence procedures for the in-line tube are usually simpler than the procedure for conventional picture tubes. When making convergence adjustments on an in-line picture tube, refer to the service data unless you are completely familiar with the convergence procedure for the receiver being serviced.

Fig. 10-6 shows the static convergence adjustments for one type of in-line picture tube. Note that there are no green static convergence adjustments. However, since both the blue and green beams can be moved horizontally and vertically, it is possible to converge all three beams without moving the green beam.

PURITY ADJUSTMENTS

It has been pointed out that in order for the three-beam tube to reproduce the hues of an image properly, each beam must strike only its respective set of phosphor dots. When this condition is achieved, color purity will be at its best. If any one of the beams happens to strike some phosphor

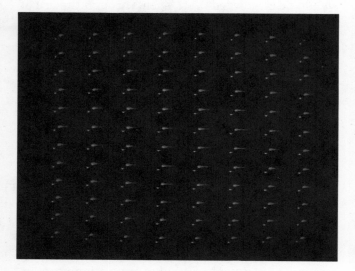

Fig. 10-5. Blue dots displaced vertically by misadjustment of the blue positioning magnet.

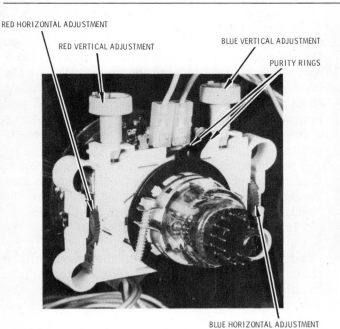

RED HORIZONTAL ADJUSTMENT

RED VERTICAL ADJUSTMENT

BLUE VERTICAL ADJUSTMENT

PURITY RINGS

BLUE HORIZONTAL ADJUSTMENT

Fig. 10-6. Static convergence adjustments for a typical in-line color picture tube.

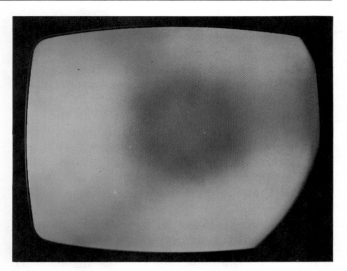

Fig. 10-7. Impure raster as a result of moving the deflection yoke back on the neck of the picture tube.

dots of the wrong color, it may be said that the color produced by this beam is impure.

Two components are adjusted or positioned to obtain optimum purity. They are the color-purity magnet and the deflection yokes. The procedure for making the purity adjustments for most receivers that use a conventional color picture tube with the phosphors arranged in dot triads is given in the following discussion.

The first step in making the color-purity adjustments is to move the yoke back on the neck of the tube until it is just against the convergence assembly. Be careful not to disturb the position of the convergence assembly. (It should be noted that some color-purity adjustment procedures specify that the deflection yoke be moved forward as far as possible.)

Since the red phosphor is usually less efficient than the green or blue phosphor, the purity adjustments should be made while observing the raster produced by the red gun. This means that the emission from the green and blue guns must be cut off during the procedure. Two methods of obtaining beam cutoff are: (1) to decrease the screen voltages by turning the screen controls to their counterclockwise positions, or (2) to ground the control grids through 100K resistors.

The screen control for the red gun should be advanced to maximum (clockwise) to obtain maximum emission from the red gun. The contrast control should be set at its counterclockwise position

or the service switch should be in the PURITY position so that no signal will be applied to the picture tube. The red beam alone is not always sufficient to illuminate the screen very brightly; therefore, it may be necessary to reduce the light in the work area.

At this point in the procedure, the center of the raster should be red as shown in Fig. 10-7. When the yoke has been moved back as described in the first step, the edges of the raster will be composed of colors other than red, indicating that the beam from the red gun is striking phosphor dots of the wrong color. The first step toward correcting this condition is to adjust the purity magnet. This device is similar to the centering magnet used in many monochrome receivers. On the 90-degree deflection tubes, the purity magnet is nearest the socket.

The purity magnet is adjusted in two ways. The strength of the magnetic field may be varied by rotating one of the rings with respect to the other, and the direction of the field may be varied by rotating both rings together. The magnetic field should not be made any stronger than is absolutely necessary. The best procedure is to begin with a very weak field and to increase the strength gradually until the best results are obtained. The tabs of the two rings should be placed adjacent to each other, and both rings should be rotated together while the effect upon the raster is noted. If there is no appreciable change in the raster, the rings are producing a minimum field. If the raster changes to a considerable degree, the rings are set incorrectly; therefore; one ring should be rotated 180 degrees with respect to the other.

The separation between the tabs should then be gradually increased, and the device should be rotated. The purpose of these adjustments is to obtain a pure red area in the center of the screen.

After the purity device has been properly adjusted, the next step is to slide the yoke forward while the raster is being observed. The red area in the center of the screen should increase in size. The yoke will be positioned properly when the red area covers the entire screen (Fig. 10-8). The purity magnet may require a slight readjustment in order that optimum color purity may be obtained.

The following is the order in which adjustments should be made for obtaining color purity:

1. Position the deflection yoke as far back as possible.
2. Cut off the beams from the green and blue guns, and set the contrast control to its counterclockwise position.
3. Advance the screen control for the red gun, and advance the brightness control.
4. Adjust the purity magnet for an area of pure red illumination at the center of the screen.
5. Move the yoke forward so that the red area covers the entire screen, and secure the yoke in this position.
6. Readjust the purity magnet for optimum purity.
7. Restore the beam currents from the green and blue guns, and adjust the screen controls for a gray screen.

The degree of purity that can be obtained on a color receiver will determine the quality of both the color and the black-and-white reproduction.

An improvement in the final purity adjustments can be made by observing the beam landing on the individual phosphor dots. A low-power microscope can be used to observe the position of the beam on the phosphor dot. The view of the dots through a microscope will be reversed and will appear as shown in Fig. 10-9A. This figure shows a normal or correct position for the beams to strike the phosphor dots.

The equilateral triangle formed by the beams is a result of proper adjustment of the static convergence magnets. These adjustments can be checked by observing the center of the screen.

Fig. 10-9B shows the direction in which the beams will move when the purity magnet is rotated. The radius of the circular movement will be changed by the strength of the field. With the tabs together the movement will be negligible; the movement will increase as the field strength increases.

Changing the strength of the purity magnet produces a straight line movement as shown in Fig. 10-9C. The direction of the movement will be determined by the rotational position of the purity magnet at the time the field is changed.

Fig. 10-9D shows the movement of the beams caused by moving the deflection yoke forward or backward on the neck of the tube. The beams are moved radially from or toward the center of the picture tube screen.

The purity adjustment procedure for in-line picture tubes is similar to the procedure just outlined

Fig. 10-8. Pure red raster.

(A) Normal landings (centered).

(B) Circular movement caused by rotation of the purity magnet.

(C) Straight-line movement caused by changing the strength of the purity magnet.

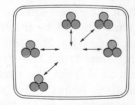

(D) Radical movement caused by moving the deflection yoke.

Fig. 10-9. Beam landings and movements produced by making purity adjustments. The dots are shown reversed as they would appear in a microscope.

for conventional picture tubes. However, when the deflection yoke is moved back, a wide vertical band usually appears in the center area of the screen. Fig. 10-10 represents the screen of an in-line picture tube with the deflection yoke moved back for the purity adjustment. The purity magnet for most in-line picture tubes is adjusted until the band is of uniform color and centered on the screen. For the example shown in Fig. 10-10, the purity is adjusted with the red beam turned on. After the band is properly centered, the yoke is moved forward to obtain a uniform red screen, and then the yoke is tightened.

The purity adjustment procedure for the Sony Trinitron is a little different than the example just given. The deflection yoke is loosened and moved forward until it is against the bell of the picture tube. The red and blue background controls are turned fully counterclockwise. The green background control is then turned fully clockwise to produce a wide green vertical band in the center area of the screen. Then the purity magnet is adjusted until the green band is centered on the screen. The deflction yoke is slid back on the neck of the picture tube to produce a uniform green raster. After the red and blue rasters have been checked for correct purity, the deflection yoke is clamped in place.

DYNAMIC CONVERGENCE ADJUSTMENTS

First, we will discuss the dynamic convergence adjustments for conventional color picture tubes using the triad dot arrangement for the phosphor screen. Both static and dynamic forces are re-

quired to maintain beam convergence during the scanning periods. The four static convergence adjustments (blue, red, and green beam-positioning magnets and the blue-lateral correction magnet) are also adjusted during the dynamic convergence procedure. There are usually 12 dynamic convergence controls used with conventional color picture tubes. In some instances there may be more or less controls, but for this discussion we will consider the usual 12 dynamic convergence controls. These controls include six vertical convergence controls and six horizontal convergence controls.

The 12 dynamic convergence controls in the approximate order of adjustment are:
The vertical convergence controls.

1. R-G vertical lines, bottom.
2. R-G vertical lines, top.
3. R-G horizontal lines, bottom.
4. R-G horizontal lines, top.
5. Blue horizontal lines, bottom.
6. Blue horizontal lines, top.

The horizontal convergence controls.

7. R-G vertical lines, right.
8. R-G vertical lines, left.
9. R-G horizontal lines, right.
10. R-G horizontal lines, left.
11. Blue horizontal lines, right.
12. Blue horizontal lines, left.

The dynamic convergence of a color receiver is usually performed in an order specified by the manufacturer. In many cases, a label inside the receiver identifies the controls and the order in which they are adjusted.

Chart 10-1 provides a dynamic convergence procedure that can be followed when the controls are identifiable according to the names used on the chart. A pattern generator must be used to produce a dot or cross-hatch pattern that is visible on the screen. Most pattern generators provide both cross-hatch and dot outputs. It is usually advantageous to use both patterns during the dynamic convergence procedure.

It may be easier to converge the red and green beams with the blue gun cut off. This can usually be accomplished by turning the blue screen control fully counterclockwise or by connecting a 100K resistor from the control grid of the blue gun to ground. Of course, it will be necessary to reactivate the blue gun in order to converge the blue beam.

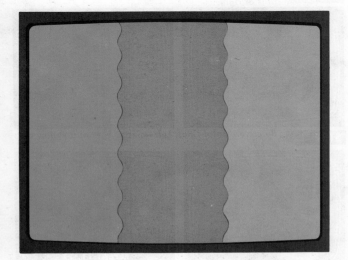

Fig. 10-10. Appearance of raster on in-line picture tube with deflection yoke moved back for purity adjustment.

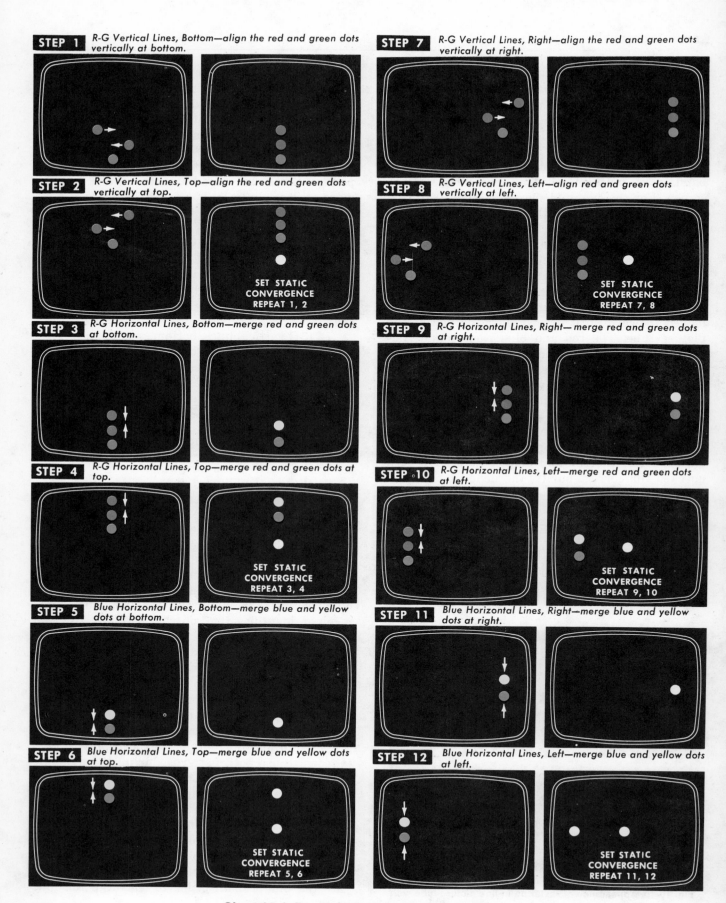

STEP 1 R-G Vertical Lines, Bottom—align the red and green dots vertically at bottom.

STEP 2 R-G Vertical Lines, Top—align the red and green dots vertically at top.

SET STATIC CONVERGENCE REPEAT 1, 2

STEP 3 R-G Horizontal Lines, Bottom—merge red and green dots at bottom.

STEP 4 R-G Horizontal Lines, Top—merge red and green dots at top.

SET STATIC CONVERGENCE REPEAT 3, 4

STEP 5 Blue Horizontal Lines, Bottom—merge blue and yellow dots at bottom.

STEP 6 Blue Horizontal Lines, Top—merge blue and yellow dots at top.

SET STATIC CONVERGENCE REPEAT 5, 6

STEP 7 R-G Vertical Lines, Right—align the red and green dots vertically at right.

STEP 8 R-G Vertical Lines, Left—align red and green dots vertically at left.

SET STATIC CONVERGENCE REPEAT 7, 8

STEP 9 R-G Horizontal Lines, Right— merge red and green dots at right.

STEP 10 R-G Horizontal Lines, Left—merge red and green dots at left.

SET STATIC CONVERGENCE REPEAT 9, 10

STEP 11 Blue Horizontal Lines, Right—merge blue and yellow dots at right.

STEP 12 Blue Horizontal Lines, Left—merge blue and yellow dots at left.

SET STATIC CONVERGENCE REPEAT 11, 12

Chart 10-1. Dynamic Convergence Procedure

At regular intervals during the convergence procedure, it is usually necessary to repeat the static convergence for the center area of the picture tube. After the center area has been converged, the two preceding steps in Chart 10-1 should be repeated.

After the convergence procedure has been completed, there may be small areas of the picture that look as though they might be improved. Although it is usually not a standard part of the convergence procedure, extremely accurate convergence can be obtained by repeating all of the steps in the convergence procedure, including the purity adjustments.

The vertical dynamic-convergence procedure is very similar for most receivers because there is a minimum of interaction between the controls. The horizontal dynamic-convergence procedure is generally more critical because there is more interaction between the controls. This interaction between controls is not identical for all receivers, and this fact accounts for the differences in the convergence procedures employed in different color receivers.

If a receiver is reasonably well converged, do not change the setting of any controls at random, but proceed with a complete convergence alignment. This procedure will, in effect, refine the present adjustments and reduce the time needed to converge the receiver.

The dynamic convergence adjustments for in-line or single-gun color picture tubes are quite different from the procedure just discussed. In fact, some color receivers employing in-line picture tubes have no dynamic convergence controls. Such is the case for RCA color receivers employing the precision in-line picture tube described in the previous chapter. Other color receivers employing in-line picture tubes have taps on the convergence assembly that may be used to change the shape of the currents through the dynamic convergence electromagnets.

Those color receivers with in-line picture tubes that do employ dynamic convergence controls usually have only two or three such controls. Due to the wide variation in dynamic convergence adjustments for in-line color picture tubes, the technician should refer to the service data for the correct dynamic convergence procedure.

ADJUSTING THE GRAY SCALE

The gray-scale adjustments represent the final steps toward completing the receiver adjustments associated with the picture tube. The term "gray scale" pertains to the various luminance values from black to white. The different values in the gray scale can be reproduced by changing the intensity of light. As the intensity is increased, the values of gray will become lighter; and as the intensity is decreased, the values of gray will become darker.

In order for the color picture tube to reproduce the proper gray scale, the light emissions from the three color phosphors must be equal. Because of the difference in the efficiencies of the phosphors, the intensities of the three beams will not be equal when the light emissions from the three phosphors are equal. The phosphor efficiencies in one particular tube are such that the red gun contributes 42 percent, the green gun 30 percent, and the blue gun 28 percent of the total ultor current when white light is produced. These percentages will vary slightly in different types and makes of color picture tubes because of differences in the phosphors used.

The proper relationship between the beam currents must be maintained at all times during the reproduction of a monochrome image in order for the various luminance levels to appear as values of gray. When the proper operating voltages for each gun are established, the light outputs from the three phosphors will remain equal as the bias levels of the guns are varied simultaneously. The color picture tube can reproduce monochrome images only under these conditions. Once these conditions are established for a monochrome image, the operating voltages for the guns will also be proper for the reproduction of color images.

The schematic diagram in Fig. 10-11 shows the circuits used to determine the operating voltages for each gun in the color picture tube. Note that there are three screen controls, three drive controls, and a picture-tube bias control. The drive controls are used to individually adjust the static bias levels of their associated guns. Simultaneous variation of the bias levels of all three guns is accomplished by adjustment of the picture-tube bias control. Each screen control is used to adjust the voltage at the screen-grid of its associated gun.

Because of the various circuits used, the procedure for adjusting the gray scale varies for different receivers. A receiver that combines the luminance signal with the color-difference signals before applying them to the picture tube usually incorporates video gain controls. These controls are used in the matrix section so that the three signals applied to the picture tube will receive the

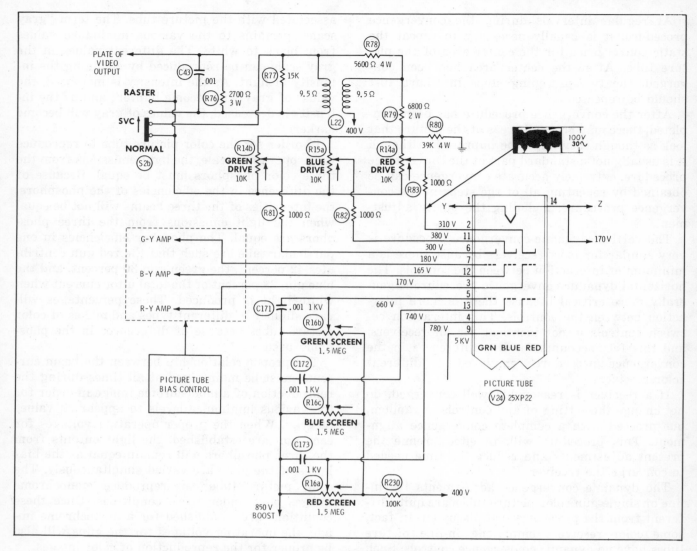

Fig. 10-11. Control circuits used to adjust the picture-tube voltages.

proper amplification. In a receiver of this type, the video gain controls must be adjusted so that proper reproduction of the gray scale will be obtained.

Some receivers use the picture tube to combine the luminance signal with the color-difference signals. The luminance signal is applied directly to the driven elements of each gun, and no video gain controls are involved in the procedure for adjusting the gray scale.

Although there are some variations in procedure, the screen controls are usually adjusted at low brightness. The video gain controls, if used, are adjusted for a black-and-white picture when the brightness and contrast controls are set for normal operation. The drive controls are usually adjusted with the brightness control set for low-brightness reproduction. When the gray scale has

been adjusted properly, monochrome reproduction should be maintained as the brightness control is varied over its normal range.

A procedure used in making the adjustments to obtain the proper gray scale on a color chassis using a conventional picture tube is outlined as follows:

Tune in a black-and-white picture or a color picture with the color control set at minimum. Set the brightness and contrast controls to midrange. Turn the picture-tube bias control to its minimum (counterclockwise) position. Turn the three screen controls fully counterclockwise. Move the service switch to the SERVICE position. Advance the screen controls one at a time until each produces a barely visible line.

If one or more of the screen controls fail to produce a line, leave that control at maximum and advance the picture-tube bias control until a barely visible line appears. Then, adjust the other two screen controls to produce a barely visible line. Move the service switch to the RASTER position. Set the brightness control to maximum and adjust the three drive controls for a gray raster. Return the service switch to the NORMAL position.

Set the brightness and contrast controls for a normal picture. Check the tracking of the three guns through the usable range of the brightness control. Any deviation from gray should be corrected by adjusting the drive controls. Do not readjust the screen controls or the picture-tube bias control.

Some older color receivers have only two drive controls—usually blue and green. Also, the controls may be called *background* controls rather than drive controls. However, the adjustment is usually the same as for the drive controls in the example just described.

Since most in-line picture tubes have a single screen grid, they have only one screen control. Therefore, the color-tracking adjustment will be somewhat different from the procedure described for the conventional color picture tube. Adjusting the gray scale for these receivers usually involves adjusting a screen control and three drive or bias controls. Always follow the instructions given in the service data for these receivers.

QUESTIONS

1. A picture tube employing electromagnetic convergence uses four components to obtain static convergence. What are they?
2. What procedure should be followed before the purity and final convergence adjustments are made?
3. What is the purpose of setting different operating voltages on the three picture-tube guns?
4. What is the advantage of using a low-power microscope during picture-tube convergence?
5. How many controls or adjustments are usually provided for complete convergence of the conventional color picture tube?
6. Referring to question five, how many controls are used for
   (a) Static convergence?
   (b) Vertical convergence?
   (c) Horizontal convergence?

Chapter 11

Troubleshooting the Color Receiver

The troubleshooting procedures for a color receiver are similar to those for a monochrome receiver. First, the picture on the screen is analyzed so that a diagnosis of the problem can be made. Then, it is decided in what sections of the receiver the cause of the trouble is most likely to be located. After this decision has been made, the cause of the trouble can be found more rapidly. It is very important to thoroughly analyze the picture produced by a color receiver before starting the troubleshooting procedure. Because there are more sections in a color receiver, much time can be wasted if the picture is analyzed incorrectly.

After it has been decided in what section or sections of a tube-type receiver the circuit defect is located, the first step in servicing procedures is to check the tubes involved. When the tubes are eliminated as possible trouble sources, the signal is traced through the circuit to isolate the defect to a specific stage. This is the procedure that will be followed in this discussion of troubleshooting.

The troubleshooting procedures for solid-state color television receivers are somewhat different from the procedures associated with tube-type receivers. Since most solid-state devices are soldered into the circuit, they usually cannot be unplugged and tested as tubes can. Therefore, when troubleshooting solid-state receivers, it is usually more practical to locate a defective component by signal tracing or by making voltage and resistance measurements. Of course, if it is fairly obvious that a trouble symptom is caused by a particular component, then it would be practical to remove that component and test it. Many solid-state receivers now use circuit modules, making it easy to substitute a complete circuit during the troubleshooting procedure.

It will be assumed that a color-bar generator is available. When trying to find a trouble in a color receiver, it is necessary to have a color signal—either a transmitted signal or one from a color-bar generator. If color broadcasts are not transmitted at all times, it will be necessary to have a color-bar generator to test all sections of the color receiver.

Troubles that occur in a color receiver fall into two main categories, monochrome and color. Monochrome troubles are those that cause improper reproduction of a black-and-white picture. The troubles that affect monochrome operation will also cause a faulty color picture. The category of color troubles can be broken down into conditions of no color, wrong colors, or loss of color synchronization. The condition of no color covers all troubles that cause the complete loss of color in the reproduced picture. In the case of wrong colors, the receiver is producing colors, but they are not the correct hue, saturation, or brightness. Loss of color synchronization is signified by the fact that the colors are present, but horizontal or diagonal stripes of variegated colors, either stationary or in motion, appear on the screen.

CHECKING MONOCHROME OPERATION

The first thing to do when beginning the troubleshooting procedure for a color receiver is to tune in a transmitted signal with the color control set at minimum and to check the operation of the receiver. If the results are good for monochrome, it is known that all the circuits associated with the reproduction of a monochrome picture are functioning properly. This means that the luminance signal is arriving at the picture tube correctly. If there is anything wrong with the monochrome picture, the color picture also would not be correct because the luminance signal is combined with the color signals; therefore, any monochrome troubles that are present must be eliminated before the color operation of the receiver is considered in the troubleshooting procedure.

Servicing the monochrome section of the color receiver is the same as it is for a conventional monochrome receiver. The sections of concern are those through which the luminance signal passes. These sections are the rf, i-f, and video-detector stages, and the luminance channel. The luminance channel in most color receivers consists of two stages of video amplification, and both the luminance and chrominance signals pass through the first video-amplifier stage. At the output of the luminance channel, the luminance signal is applied either to a matrix section or directly to the picture tube.

Loss of Monochrome Picture

If the receiver does not produce a monochrome picture when receiving a signal, the cause of the trouble can be located somewhere between the input of the receiver and the output of the luminance channel. Follow the same procedure that is used when troubleshooting for loss of the picture in a monochrome receiver. Since a color receiver is under consideration, the color-bar generator can be used to advantage in isolating the stage or stages in which the trouble exists.

Connect the rf output of the color-bar generator to the antenna terminals of the receiver. If color appears on the screen, it can be assumed that the stages up to the point where the chrominance signal is separated from the composite video signal are operating properly. This means that the cause of the trouble is somewhere between the stage in which the chrominance takeoff point is located and the output of the luminance channel. The color bars on the screen would have improper brightness levels since there would be no output from the luminance channel.

If the receiver has a luminance channel similar to that shown in Fig. 11-1, the circuit between the output of the first video amplifier and the plate of the video output tube should be checked. The second video amplifier should be substituted first. A defective video output tube can also cause a loss of the luminance signal. However, a defective video output tube usually cuts off the picture tube, causing a loss of raster. If tube replacement does not eliminate the trouble, an oscilloscope should be used to trace the luminance signal through the circuit. After it has been determined just where the trouble is located in the circuit, the defective component can be isolated through voltage and resistance checks.

If color does not appear when the rf output of the color-bar generator is connected to the antenna terminals of the receiver, the cause of the trouble is located between the receiver input and the point where the chrominance signal is separated from the composite video signal. If the conventional methods of troubleshooting are followed, the defective tube or component can be found.

It can be seen from the foregoing discussion that the color-bar generator can be very useful even when troubleshooting the loss of the monochrome picture. Time can be saved if it is known that the color signal is able to pass through the circuits that are also common to the luminance signal.

If the receiver has a luminance channel similar to the one illustrated in Fig. 11-2, normal signal tracing procedures should be used to locate the defective component.

Improper Gray Scale

Certain troubles will affect the ability of a color receiver to reproduce proper shades of gray. A trouble of this type is indicated if a monochrome signal is reproduced in shades of a particular color instead of gray. In some cases, the image may be reproduced in shades of a primary color; in other cases, the image may be reproduced in shades of a complementary color.

If a color receiver is unable to reproduce shades of gray, the circuits that control the voltages applied to the elements of the picture tube are not adjusted correctly, there is a defective component in one of these circuits, or one of the guns in the picture tube is defective. Assuming that an attempt to adjust the gray scale is of no avail, let us examine some typical circuits and determine the possible causes for the reproduction of a monochrome picture in a color shade instead of the proper shades of gray.

Fig. 11-3 shows the matrix section and the circuits associated with the picture tube in a tube-type color receiver. The luminance signal is applied to the cathodes, and the color-difference signals are applied to the control grids of the picture tube. Note that the plates of the color-difference amplifiers are dc-coupled to the control grids of the picture tube. A change in the conduction of one of these amplifiers will change the voltage on the control grid of the associated electron gun, and the reproduction of the gray scale will consequently be affected.

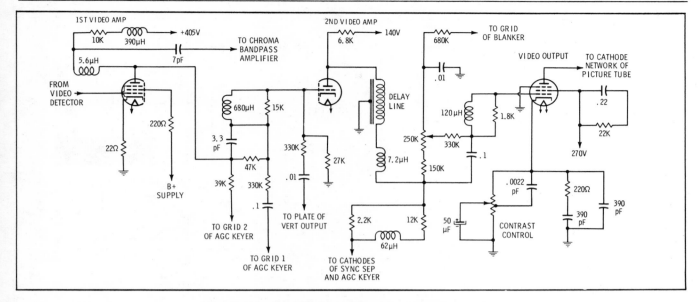

Fig. 11-1. Luminance channel in a tube-type color receiver.

The first step toward correcting such a trouble in a tube-type receiver is to check the amplifier tube associated with the deficient or predominant color. If replacement of this tube does not cause a considerable change in the image, the voltages on the cathode, control grid, and screen grid of the associated electron gun should be measured. One method of checking these voltages is to measure the voltage present at the socket of the picture tube. Caution should be observed when measuring voltages at the socket because of the presence of high voltage on some of the pins.

If the voltage on one of the elements of the picture tube is not within tolerance, the associated circuit should be checked for a defect. It is also possible for a defective electron gun to cause the voltages at the elements to be incorrect. The voltages at the socket can be measured after the socket has been removed from the tube base. If the voltages then appear to be normal, the associated electron gun may be defective.

The circuit shown in Fig. 11-4 is from a solid-state color receiver. This circuit differs greatly from the conventional vacuum-tube types. The

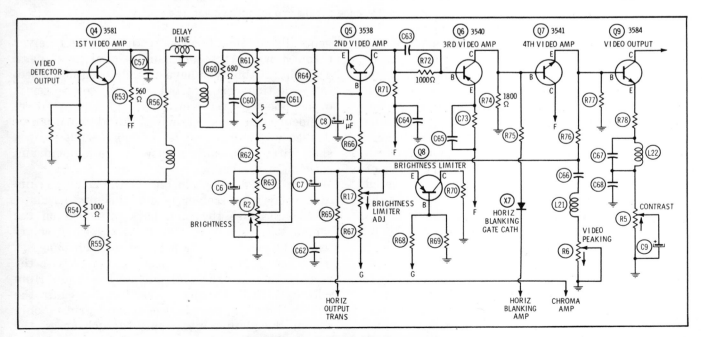

Fig. 11-2. Luminance channel in a solid-state color receiver.

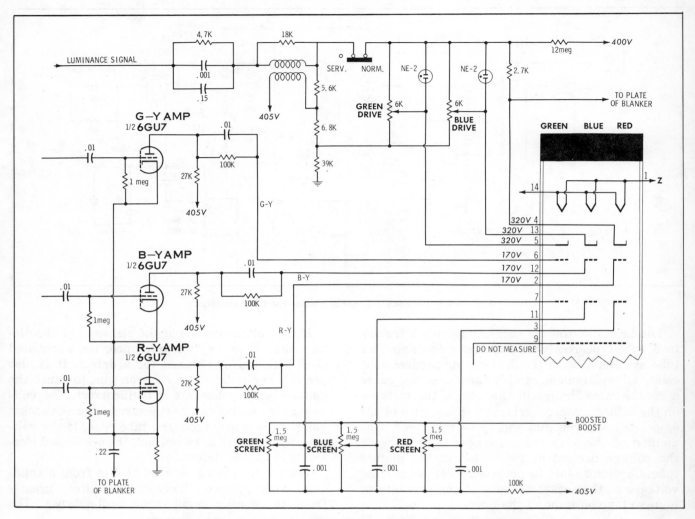

Fig. 11-3. Luminance and color-difference circuits in a tube-type color receiver.

luminance signal is coupled through each of the three separate color demodulators and passes through three separate and parallel channels to the picture tube. This luminance signal will produce a monochrome picture when no color is being received. When a color signal is being received, the luminance signal combines with the demodulated color signals at the demodulator outputs to produce red, green, and blue video signals. Each of these three video channels is coupled to its related picture-tube cathode. This circuit eliminates the necessity of matrixing color difference and brightness signals at the various grids and cathodes of the crt.

Effect of Hum on Monochrome Pictures

Normally, the color circuits do not contribute any signals to the picture tube during the reproduction of a monochrome scene. For this reason, the cause of improper monochrome operation is generally confined to those sections of the receiver through which the luminance signal must pass. Troubles of this type have already been described.

In addition, certain troubles in the color circuits of a tube-type receiver may occasionally affect monochrome reproduction. One such trouble can be defined as "hum," and is commonly caused by a short circuit or leakage between the filament and cathode of a tube.

Hum that develops in one of the color circuits of a tube-type receiver is not difficult to isolate. This is because the hum bars produced on the screen will usually appear in two specific colors when the receiver is tuned to a monochrome signal. These colors will be those associated with the receiver section that introduces the hum. Hum may develop in the matrix section and cause the modulation to appear at the control grid of only one of the picture-tube guns. In such cases, the hues produced on the screen will be the primary

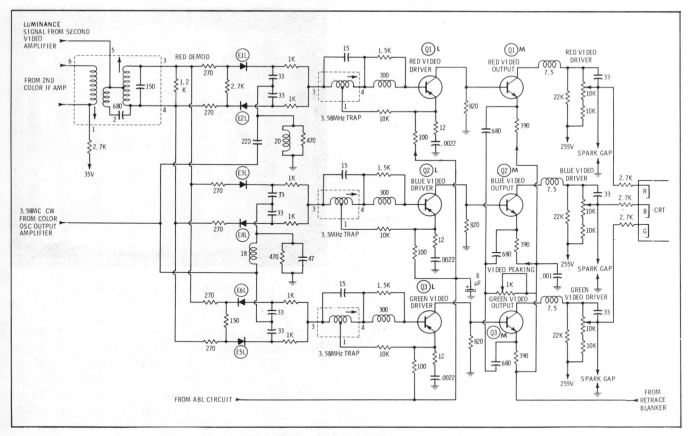

Fig. 11-4. Video output section of a solid-state color receiver.

and complementary colors associated with the gun receiving the hum modulation. Hum that modulates the red gun will cause red and cyan bars to appear on the screen.

When 60-Hz hum is caused by a defective tube, the trouble can be readily corrected. For instance, if the colors that appear on the screen during the reproduction of a monochrome picture indicate that the hum originates in the R − Y amplifier, the tubes in this section (including the demodulator) should be substituted one at a time until the hum disappears. If this does not correct the trouble, the chassis should be removed and an oscilloscope used to trace the hum in the conventional manner.

Sometimes the hum modulation may be very weak, and the colors contaminate the monochrome picture will be pale. This may make it difficult to determine where the hum originates, because the hues produced in one circuit may closely resemble those produced in another circuit. In such cases, it may be necessary to check the bandpass amplifiers, demodulator tubes, and the color-difference amplifiers.

The color-phase diagram, shown in Fig. 11-5, will be of considerable aid in determining which

part of the circuit is defective. For example, red and cyan are 180 degrees, out of phase with each other, and when these colors are affected, the trouble is probably in the R −Y circuits. In the case of a receiver employing I and Q demodulation, the trouble would be located in the I channel, since this vector is nearest the red vector.

THE COLOR-BAR GENERATOR

Color-bar generators provide various chroma signals that can be used for the adjustment and troubleshooting of color receivers. The most popular color generator is the keyed-rainbow generator which produces ten bars that vary in hue from yellow-orange on the left to green on the right. These color bars are produced by the offset-carrier method; that is, a crystal-controlled oscillator (3.563795 MHz) beats against the color set's 3.579545-MHz oscillator. Since the beat difference is 15,750 Hz, there is a complete 360-degree phase shift for each horizontal scanning line, which yields a rainbow of colors. These colors are then "keyed" into color stripes or bars so that they can be easily identified; at the same time, the keying

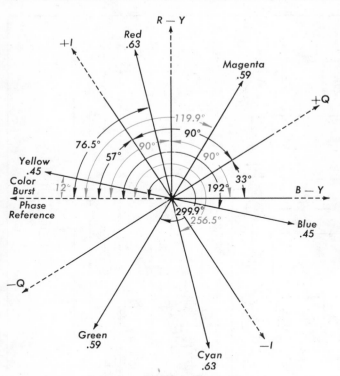

Fig. 11-5. Color-phase diagram.

Fig. 11-7. Normal pattern produced by keyed-rainbow
color-bar generator.

process modulates the color signal so that it can be traced through the circuit with a scope. The color bars are presented in the order shown in Fig. 11-6. The color-bar pattern from a keyed-rainbow generator is shown in Fig. 11-7.

RF AND I-F SIGNAL TESTS

When loss of color is encountered, a color-bar generator can be used to determine in which section of the receiver the color is being lost. After

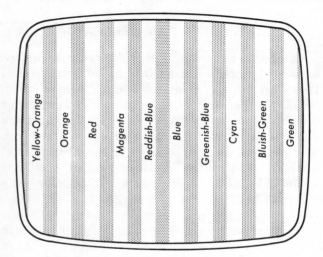

Fig. 11-6. Sequence of color bars obtained from a
keyed-rainbow color generator.

the receiver has been checked for proper monochrome operation, connect the output of the color-bar generator across the antenna terminals of the receiver. While the rf signal is being applied to the receiver, the color control should be turned to its nearly maximum position and the fine-tuning control should be rotated throughout its range. The color control is adjusted to its nearly maximum position because the receiver could be producing color, but the color might be of such a low amplitude that it would not appear on the screen unless the color control were turned up. In some color receivers, loss of color can also be caused by misadjustment of the hue, or tint, control. Therefore, this control should be rotated through its range.

If the receiver is in the home during this test and if the colors are correctly reproduced when the rf signal from the color-bar generator is being used, the loss of color may have been caused by misadjustment of the color control, the hue control, or the fine-tuning control. There is also a possibility that the antenna may not be properly oriented or that one of the lead-in wires may be broken. These troubles may be present without seriously affecting monochrome operation. When a composite color signal is being received, however, the chrominance portion of the signal reaching the antenna terminals might be attenuated to such an extent that color reproduction would be affected.

If color was not reproduced when the rf color-bar signal was applied to the receiver, the video signal from the color-bar generator should be injected at the video input of the receiver. In most cases, the generator can be connected to the input

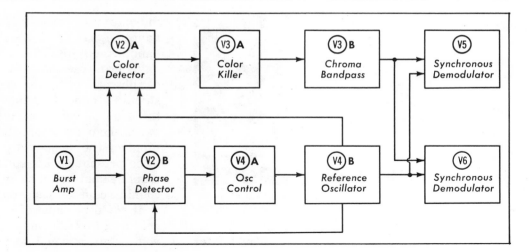

Fig. 11-8. Block diagram of the color section of a tube-type color receiver.

of the video section of the receiver without the need for removal of the chassis from the cabinet.

If color appears, the color circuits cannot be the cause of the trouble. Since color was not obtained when the rf signal was used but is obtained when the video signal is used, the color signal is not able to pass through the rf or i-f sections. Whenever this is the cause, the alignment of the rf and i-f sections should be checked. The response of these sections must be broad and flat in order for the chrominance signal to pass. The response could be incorrect even though no serious effect might be noticed in the monochrome picture. A color receiver that has been operating normally will not usually require realignment unless a tuned circuit is misaligned or a component in a tuned circuit fails. The replacement of a defective tube, capacitor, or resistor will generally return the receiver to normal operation; the i-f bandpass can then be checked to be sure it is within the desired limits.

SIGNALS IN THE COLOR SECTION

Let us now consider the procedure to follow if color is not obtained when the video signal is applied to the video circuits. This indicates that the source of the trouble is not in the rf and i-f sections but somewhere in the color circuits. The color circuits that are to be investigated in a tube-type receiver are those shown in Fig. 11-8.

The tubes that should be replaced are in the chroma-bandpass amplifier, the color afc, and the color killer circuits. It is usually not necessary to replace the color-demodulator tubes because both must be defective in order to cause a complete loss of color. There are nine stages in the color section of the receiver under discussion, but because dual-purpose tubes are used, it is necessary to check

only five tubes. The bandpass amplifier and the color killer are contained in one tube envelope, the reference oscillator and oscillator control are in another, and the phase detector and color detector are in another envelope. Most late-model tube-type color receivers employ semiconductor diodes for the phase detector and the color-killer detector.

If color is not restored after the tubes have been changed, the cause of the trouble is the failure of a component or components within the circuits under consideration.

Signals at the Demodulators

In order for the color demodulators to operate properly, the chrominance signal and the cw reference signals must be present at the input of these stages. If either the chrominance signal or the reference signals are missing, no color will be reproduced. Therefore, the most logical place to start looking for the cause of the trouble is at the inputs of the color-demodulator stages.

Connect a color-bar generator to the antenna input terminals. Check for the presence of the cw

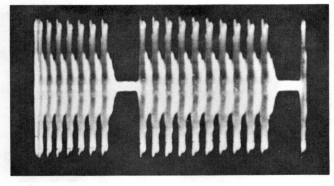

Fig. 11-9. Waveform of the chrominance signal from a color-bar generator as it appears at the input of the demodulators.

Fig. 11-10. Waveform of the cw reference signal at the input to the demodulators.

and the chrominance signals at the demodulator inputs. The chrominance signal should appear like the one shown in Fig. 11-9. The cw reference signals should appear like that shown in Fig. 11-10.

If the chrominance signal and the cw reference signals are found to be present when an oscilloscope is being used, the trouble must be in the demodulator stages. The complete loss of color can be attributed to the demodulator circuits only if there is a portion of the circuit that is common to both demodulators or if both tubes should go bad at the same time (which is very unlikely). For example, if the screen of each demodulator is returned to a voltage source through a common resistor and this resistor were to open, no voltage would be supplied to the screens and both demodulators would be inoperative.

Let us consider a receiver in which the reference signals are present at the color demodulators, but in which the chrominance signal is absent. This

indicates that something is preventing the chrominance signal from passing through the chroma bandpass circuit. To locate the point where the signal is lost, trace back through this circuit with the oscilloscope until the signal is found. When the signal is found, it is then known that the cause of the trouble exists somewhere in the circuit between the point where the signal is present and the input of the demodulators. If the signal is found to be present on the control grid of the bandpass amplifier but not at the plate, the loss of the signal could be caused by a drop in plate or screen voltage or it could be caused by the fact that the stage is held at cutoff by the action of the color killer. A voltage check could be used to determine quickly which was the case. If the control grid has a high negative potential, it should then be determined why the color killer is biasing the bandpass amplifier to cutoff.

The color killer is normally held at cutoff by a negative potential which is applied to its grid. This potential is developed by one of the phase detectors and is always present as long as there is a color-burst signal. When the color burst is absent at the phase detectors, the color killer is allowed to conduct and to bias the bandpass amplifier to cutoff.

To determine whether or not the color burst is arriving at the phase detectors, check for the presence of the signal by using an oscilloscope. The color-burst signals applied to the phase detectors should appear like those shown in Fig. 11-11. If burst signals are present, check the color-killer circuit. In the circuit of Fig. 11-11, the grid of the

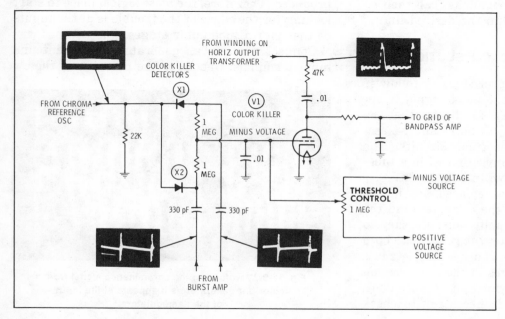

Fig. 11-11. Color-killer circuit.

color killer would go to zero potential if the bypass capacitor at the grid were to become shorted. The color killer would therefore conduct, and the loss of color would result.

If the color burst is not present, go back and check through the burst-amplifier circuit with the oscilloscope. The color burst should appear at the plate of the burst amplifier. The signal at the input of the burst amplifier should be the composite color signal which would appear like that shown in Fig. 11-12. If this signal is present at the input but the color burst is not present at the output, check the circuit associated with the burst amplifier.

The burst amplifier is caused to conduct during horizontal-retrace time either by a pulse from a burst-keyer stage or by a pulse that is taken from a winding on the horizontal-output transformer. Since the color burst is transmitted during horizontal-retrace time, only the color burst would be amplified by the burst amplifier. If the keying pulse is lost, the burst amplifier would be in a non-conducting state at all times; consequently, no color burst would appear at the output.

The chrominance signal can be lost in the chroma-bandpass circuit anywhere between the chrominance takeoff point and the input of the demodulators. The bandpass amplifier could be biased to cutoff by the color killer because of the loss of the color burst. The color burst can be lost anywhere between the burst-takeoff point and the input of the phase detectors. The color burst will not reach the phase detectors if no keying pulse is applied to the burst amplifier. If a color killer is not employed to disable the bandpass amplifier during the absence of the color burst, it would only be necessary to check for the loss of the chrominance signal in the chroma-bandpass circuit.

Chrominance Signal Present But Reference Signals Absent

If it is found that the chrominance signal is present but both reference signals are absent when the signals at the inputs of the color demodulators are checked, the chroma-bandpass circuit should be disregarded and the color-sync circuit should be checked. In order to have complete loss of color, the reference signals at the input to both demodulators must be absent. If one reference signal is present, colors would be reproduced but they would be incorrect. The reference signal that should be present at the demodulators is shown in Fig. 11-10.

There are two possible reasons why the reference signals are missing at the demodulators.

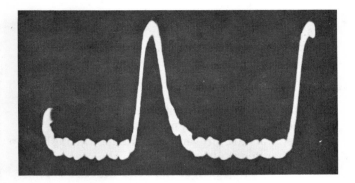

Fig. 11-12. Waveform at the grid of the burst amplifier.

Either the reference signal is not being generated by the oscillator, or the signal is being lost between the oscillator and input to the demodulators.

Loss of the oscillator output signal can result from a defective tube, detuning of the 3.58-MHz output coil, or a bad crystal. With either a detuned tank circuit or a bad crystal, the plate voltage of the oscillator drops below the value it should be. Substitution is the best way to check for a faulty crystal. If the crystal is not at fault, voltage and resistance checks should then be used to locate the component that is defective.

In some instances, an apparently dead oscillator can be restarted by a simple touch-up adjustment of the reactance-tube plate coil. A trial alignment may help solve a serious sync problem by revealing that the oscillator is being thrown off frequency by a defective phase detector.

Let us summarize the steps of the troubleshooting procedure for the condition of complete loss of color. If a satisfactory monochrome picture can be produced by the color receiver, apply an rf signal from a color-bar generator to the input of the receiver. If color appears on the screen after the controls have been properly adjusted, check for an improperly oriented antenna or a broken antenna lead-in wire.

If color is not produced by the rf signal, apply the video signal to the input of the video section. The results obtained will determine whether the trouble is ahead of or after the video input. If color appears on the screen, the trouble is in the rf or i-f section. The alignment of these sections should be checked.

If color does not appear when the video signal is used, the trouble is in one of the color circuits somewhere between the point where the chrominance signal is separated from the composite color signal and the output of the demodulators. If the chassis is in the cabinet, check the tubes in the chroma-bandpass, color-sync, and color-killer circuits.

With the chassis removed from the cabinet, check the signals at the demodulators with an oscilloscope. When both the chrominance and reference signals are present, the trouble is in a circuit that is common to both demodulators.

If the chrominance signal is absent but the reference signals are present, check back through the chroma-bandpass circuit to locate the place where the signal is being lost. If the color killer is biasing the chroma-bandpass amplifier to cutoff, check the color-killer circuit or check for the absence of the color burst at the color phase detectors. The color burst can be lost anywhere between the burst take-off point and the color phase detectors. If the correct signal is present on the control grid of the burst amplifier but the color burst is not present at the plate, check for the loss of the keying pulse from the horizontal-output transformer.

If it is found that the chrominance signal is present but the reference signals are absent at the demodulators, check the 3.58-MHz oscillator to locate the point where the reference signal is being lost.

The troubleshooting procedures for loss of color in a solid-state receiver depend somewhat on the type of solid-state devices used in the circuit. When discrete transistors are employed in the color circuits, an oscilloscope can be used to trace the color signal through each stage. After determining at which stage the color signal has been lost, the defective component con usually be isolated by voltage and resistance measurements. If transistor sockets are used, suspected transistors can be removed and checked with a transistor checker.

Although color circuits employing discrete transistors generally have more stages than their tube-type counterparts, the basic function is still the same as shown in Fig. 11-8. Therefore, many of the procedures described for tube-type receivers are applicable to the solid-state sets. In some receivers, both discrete transistors and ICs are used in the color circuits. For example, an IC may be employed as the color demodulator. When a plug-in IC is used, suspected troubles in the demodulator circuit can be checked by substituting the IC. Such ICs will also be found in some hybrid receivers.

In many late-model color receivers, the color signals are processed by two or three ICs. Such a circuit is shown in Fig. 11-13. Except for the chroma detector and the 3.58-MHz amplifier, all of the active components for the color circuits are contained within the two ICs. Therefore, the most likely source for the loss of color is the ICs. Of course, when the ICs are mounted in sockets, they can be checked by substitution. If the ICs are soldered into the circuit, signal tracing can be used to determine which part of the circuit is at fault.

The first step would be to check for the chroma signal at pin 3 of the demodulator, IC602. If there is no signal present at pin 3, the next step is to look for an input signal at pin 5 of the color processor, IC600. The presence of a signal at this point would indicate that the color processor is at fault. If the voltages on the various pins of the IC are within reason, the color processor IC should be replaced. When a normal chroma signal is present on pin 3 of the color demodulator, check for the presence of the chroma reference signal on pins 6 and 7. If the reference signal is missing at both pins, the 3.58-MHz amplifier and the burst processing circuit should be checked. When the chroma signal and both of the reference signals are present at the demodulator and there is no color, IC602 is probably defective and should be replaced.

In many solid-state color receivers, the entire color circuitry is contained on one or two plug-in modules. Often, the most practical service procedure in this case is to check the module by replacing it with a known good module. If the original module is defective, it can be repaired by using normal troubleshooting procedures or it can be replaced.

WRONG COLORS

If monochrome operation is normal but improper colors are being produced during a color transmission, the trouble can be classified as reproduction of wrong colors. A color can be defined by its hue, saturation, and brightness. A change in any one of these characteristics will cause improper color reproduction.

A color receiver determines the hue represented by the chrominance signal by comparing the phase of this signal with the phases of two reference signals. If the phase relationship between any two of these signals is incorrect, the hues represented by the chrominance signal will be reproduced incorrectly. Saturation is determined by the amplitude ratio between the chrominance and luminance signals. If this ratio is altered in the receiver, color saturation will be affected.

In the following discussion, the luminance signal is considered normal. Although a serious change in the amplitude of the luminance signal would affect color saturation, the symptoms in

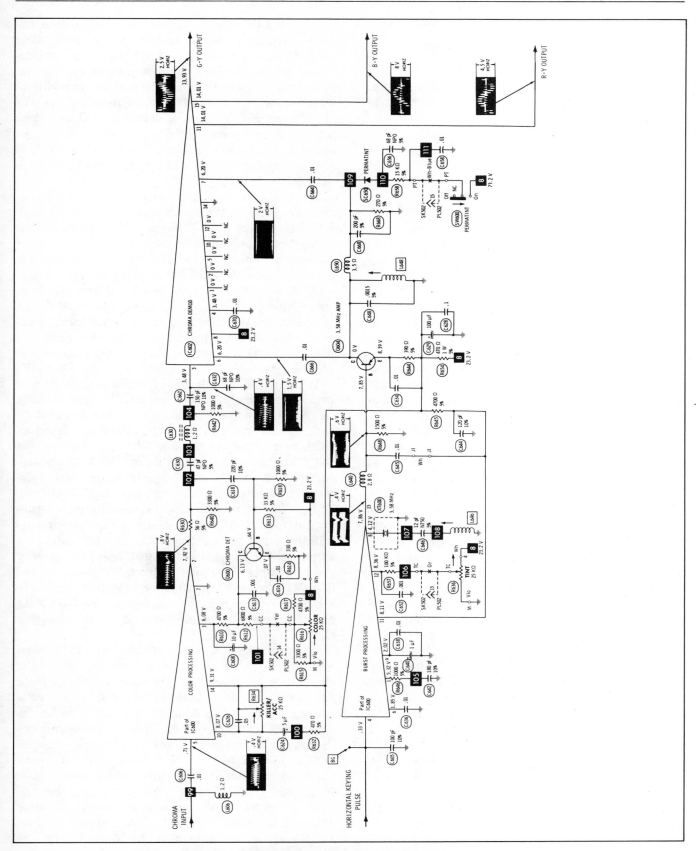

Fig. 11-13. Color section in a late-model, solid-state color receiver.

such a case would be classified as improper monochrome operation rather than wrong colors.

A number of indications that will help to isolate a trouble in a color receiver can be obtained before the back cover is removed. After making sure that the receiver operates satisfactorily when tuned to a monochrome signal, its performance should be checked when it is tuned to a color signal. A good source of such a signal is a color-bar generator. If the picture produced on the screen when the receiver is tuned to a signal from the generator is observed, several indications may be obtained to help isolate the trouble to a particular section of the receiver.

The first step toward restoring normal receiver operation is to check for a misadjustment of any controls that might be causing the trouble. The misadjustment of any one of three controls in particular may cause the hues represented by a color signal to be reproduced incorrectly. These controls are the hue (tint), saturation (color), and fine-tuning controls.

A misadjusted hue control will cause the phase relationships between the chrominance signal and the two reference signals to be incorrect. A misadjusted saturation control will cause the amplitude ratio between the chrominance and luminance signals to be incorrect. A misadjusted fine-tuning control may cause the signal to be distorted so that either or both of these troubles will occur. The setting of all three of these controls should be checked before it is assumed that the trouble is due to a defective component. If an rf signal from a color-bar generator is used during these checks, make sure that the generator and the receiver are tuned to the same channel; otherwise, some very erroneous indications may be obtained.

If normal receiver operation cannot be restored by adjustment of the controls, the tubes employed in the section suspected of causing the trouble should be checked. A good method of finding faulty tubes is to substitute one known to be good for each one in the suspected section. In the event that proper operation cannot be restored by the adjustment of controls or the replacement of tubes, the chassis will have to be removed from the cabinet for further analysis.

Wrong colors can also be caused by loss of the color-difference signals. Two such signals are usually produced by the chroma demodulation process in tube-type color receivers. Depending on the type of demodulation process used, these signals may be any two of the $B - Y$, $G - Y$, and $R - Y$ color-difference signals or they may be other sig-

nals such as X and Z. Ordinarily, when one of the color-difference signals is lost in a receiver that demodulates only two of the signals, the colors produced on the screen are those represented by the amplitude variations of the remaining color-difference signal. In a receiver that demodulates the $R - Y$ and $B - Y$ signals, values of red and cyan would be produced on the picture-tube screen in absence of the $B - Y$ signal, and values of blue and greenish-yellow would be produced in absence of the $R - Y$ signal.

In receivers that demodulate on other angles, such as X and Z, the colors produced on the screen when one of the color-difference signals is lost will depend on the demodulation angle and the type of demodulator circuit used. Most solid-state color receivers demodulate all three color-difference signals, $B - Y$, $G - Y$, and $R - Y$. When all three color-difference signals are demodulated, a defect in one of the demodulator circuits will usually affect only one of the primary colors. Of course, any other color that contains that primary color will also be affected.

When the demodulator circuits are suspected as being the cause of wrong colors, they can usually be checked by using a color-bar generator and an oscilloscope. With a color-bar pattern displayed on the screen, compare the color-difference signals at the outputs of the demodulators with the waveforms shown in the service data. If one of the demodulator outputs is missing, is low in amplitude, or has the wrong phase, the demodulator is probably defective. When all of the demodulator outputs have the wrong phase, the chroma reference signal may also be the wrong phase. If adjusting the hue control will not correct the problem, the 3.58-MHz output transformer may be incorrectly aligned or there may be a defect in the 3.58-MHz oscillator circuit or in the color phase-detector circuit.

When wrong colors are encountered in a tube-type circuit, the demodulator tubes should be checked first. If the tubes are good, check the plate and screen voltages. Missing plate or screen voltages may cause the demodulator stage to be inoperative. Next, check the circuit that produces the phase shift for the chroma reference signal. Sometimes a defect in this circuit will cause both chroma reference signals to have the same phase. As a result, the color-difference signals at the outputs of the demodulators will also have the same phase. The solid-state color circuit shown in Fig. 11-13 employs a single IC as the demodulator. In this type of circuit, the most likely source for wrong colors is a defective IC or improper chroma reference signals.

Loss of Color Saturation

One specific case in which the colors represented by a signal are reproduced incorrectly can be defined as a loss of saturation. This condition is shown by the colors in the pattern in Fig. 11-14. By comparing this photograph with the normal color-bar pattern in Fig. 11-15, it may be noted that the colors in the bars in Fig. 11-14 appear as pastels instead of saturated colors.

A pastel is produced by a mixture of white light and a fully saturated colors. Since white light contains all colors, all three of the phosphors must be emitting a percentage of the total light output during the scanning of each color bar. This means that the bars that represent the primary colors contain some amount of light from the other two primaries and that each secondary color contains some amount of light from its complementary primary. For example, the red bar normally contains only light from the red phosphor. When this bar becomes desaturated, it will also contain some light from both the green and the blue phosphors. The cyan bar normally contains only light from the green and blue phosphors. It becomes desaturated when light from the red phosphor is added.

Such a condition is caused by the fact that the signals applied to the picture tube are unable to cut off each beam at the proper time. The luminance signal is known to be normal; therefore, it may be assumed that the desaturation of colors is caused by a deficiency in the color-difference signals. It may be further assumed that the inability of these signals to cause proper beam cutoff is due to insufficient signal amplitude. Since all of the

Fig. 11-15. Color bars with normal saturation.

colors are desaturated, the signals from each demodulator section must be lacking in amplitude.

In some instances, it may be possible to remedy a condition of desaturated colors by advancing the saturation (color) control because the setting of this control determines the amount of signal applied to the demodulator stages. However, if it becomes necessary to advance this control beyond its normal setting in order to obtain sufficient color saturation, a definite trouble is indicated.

There are two possible causes for this trouble. Either the chrominance signal applied to the demodulator stages is lacking in amplitude, or the cw reference signals applied to these stages are lacking in amplitude. The stages that might cause a loss of saturation are the bandpass amplifier, the oscillator, and the burst amplifier. A weak tube in the bandpass-amplifier stage will not provide the proper amplification of the chrominance signal, and a weak tube in the oscillator stage will not provide the proper amplification of the 3.58-MHz reference signal. A weak tube in the burst-amplifier stage may not amplify the burst signal a sufficient amount to cut off the color killer completely. This would increase the bias on the bandpass amplifier and reduce the amount of amplification provided by this stage.

Chrominance Signal Lacks Amplitude

If replacement of the bandpass amplifier, the reference oscillator, or the burst amplifier does not correct the loss in saturation, the chassis should be removed and the signals at the input of the demodulator stages should be observed on an oscil-

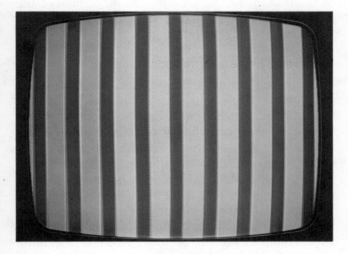

Fig. 11-14. Color bars that lack saturation.

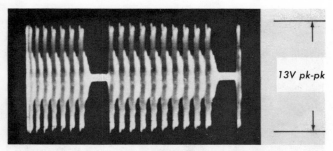

Fig. 11-16. Amplitude of the chroma signal at the input of the chroma demodulators.

loscope. The chrominance signal should appear as shown by the waveform in Fig. 11-16.

The amplitude of the waveform should be checked with the amplitude shown in the service data for the receiver. If the chrominance signal is lacking in amplitude at this point, the signal should be observed at the grid of the bandpass amplifier. The normal waveform of this signal is shown in Fig. 11-17. In some receivers, a lack of signal amplitude at this point is an indication that the 4.5-MHz trap in the cathode circuit of the first video amplifier may be detuned and is presenting a high impedance at the frequency of the chrominance signal.

If a normal signal can be observed at the input of the first video amplifier, the adjustment of the 4.5-MHz trap (if used) should be checked. If the signal at the input of the video amplifier is not normal, the passband of the video i-f section should be checked. The i-f section may be misaligned and therefore may not be providing proper amplification of the chrominance signal.

In the event that the amplitude of the chrominance signal appears to be normal at the grid of the bandpass amplifier, this stage may not be providing the proper amount of amplification. After a new tube has been tried, the voltages around this stage should be measured. A low plate or screen voltage could cause the trouble, or bias on the stage may be excessive.

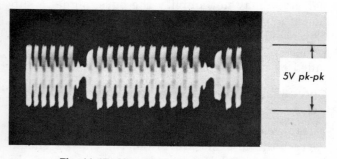

Fig. 11-17. Chroma signal at the grid of the chroma-bandpass amplifier.

The schematic in Fig. 11-11 includes the color-killer circuit. This circuit is employed to cut off the bandpass amplifier during monochrome reproduction. Similar circuits are used for this purpose in several makes of receivers. The color killer is usually held at cutoff by a negative voltage which is developed at the color phase detectors by the burst signal. A weak burst signal may not develop enough negative voltage to hold the color killer at cutoff. As a result, the color killer will conduct and a negative voltage will be developed in the plate circuit. This negative voltage will appear at the grid of the chroma-bandpass amplifier and will reduce the gain of this stage.

A negative voltage at the grid of the bandpass amplifier is an indication that the color killer may be conducting. If the burst signal observed at the phase detectors does not appear to have the proper amplitude, the voltages around the burst-amplifier stage should be measured. The alignment of the burst-amplifier transformer should be checked if no significant discrepancies are found in the voltages. Fig. 11-11 shows the waveforms of a normal burst signal at the inputs of the phase detector.

A weak chrominance signal is not as common in solid-state receivers as it is in tube-type receivers. A defect in a solid-state chroma-bandpass amplifier more often results in a complete loss of the chroma signal. However, low signal amplitude in a solid-state chroma circuit can be caused by a leaky transistor, an open emitter bypass capacitor, a misaligned chroma-bandpass transformer, or other component defects. With the signal from a color-bar generator applied to the antenna terminals, the chroma signal can be traced through the chroma-bandpass section by using an oscilloscope. By comparing the amplitude of the signal at various points in the circuit with the amplitude of the waveforms shown in the service data, the defective stage can usually be located.

Reference Signals Lack Amplitude

If the amplitude of the chrominance signal at the input of the demodulators appears to be normal, the reference signals applied to these stages should be observed. The waveform in Fig. 11-18 shows how these signals should appear. Insufficient signal amplitude indicates that the signal from the 3.58-MHz oscillator is not receiving the proper amplification. A possible cause for this is that the oscillator is not providing enough drive. A measurement of the voltages around this stage should help locate the trouble. There is a remote possibility that both the chrominance signal and

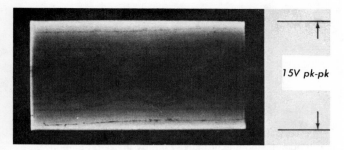

Fig. 11-18. Amplitude of the cw reference signal at the input to the chroma demodulators.

the two reference signals applied to the demodulators are normal. This would be a strong indication that the gain of each of the two demodulator stages has been reduced by an equal amount. It is very unlikely that the two tubes will become equally weak at the same time. A better guess is that some component common to these two stages is causing the trouble. In such cases, the voltages applied to the demodulators should be checked against the normal voltages shown on the schematic.

PHASE OF 3.58-MHz OSCILLATOR SIGNAL INCORRECT

A trouble may develop in a color receiver that causes it to incorrectly reproduce all of the hues represented by an incoming signal. In some cases, the hues of the colors may be changed only slightly; and in other cases, the color bars may change to completely different hues. If the hues have shifted only a small amount, an adjustment of the hue control will generally correct the trouble. If the hues have changed completely, an ad-

justment of the hue control probably will not correct the trouble. Remember, two wrongs do not make a right. Therefore, if a radical change from the normal setting of the hue control is required to restore the proper hues, there is a trouble in the color circuits, and it should be located and corrected.

Information relating to the hues represented by a transmitted signal is denoted by the phase relationship between the chrominance signal and the color burst. The hues produced on the screen of a color receiver are dependent upon the phase relationship between the chrominance signal and each of two reference signals. The source for these reference signals is a 3.58-MHz oscillator in the color-sync section. The color-sync circuitry locks the phase of the signal supplied by this oscillator in step with the phase of the color-burst signal. If this phase relationship is not accurately maintained, the hues reproduced on the screen of the receiver will be incorrect; consequently, a trouble that develops in the color-sync section can cause all of the hues to be reproduced incorrectly.

The 3.58-MHz oscillator and control circuits in a tube-type receiver are shown in Fig. 11-19. From the discussion on the theory of operation of the color-sync circuits, it may be recalled that the burst-amplifier stage is used to separate the color burst from the chrominance signal. The phase detectors compare the phase of the 3.58-MHz reference signal with the phase of the burst signal. Any phase error in the reference signal will cause a dc correction voltage to be developed. This correction voltage appears at the grid of the reactance tube and controls the current through this tube. An increase in the conduction of the reactance tube will

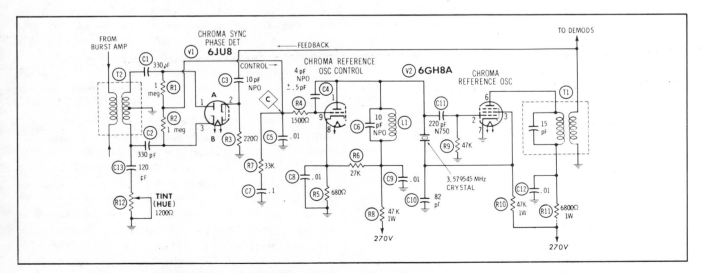

Fig. 11-19. Color-sync circuits used to control the 3.58-MHz reference signal in a tube-type color receiver.

retard the phase of the oscillator signal, and a decrease in the conduction of this tube will advance the phase of the oscillator signal.

Tube Troubles

It might appear that a weak reactance tube would result in an error in the phase of the oscillator signal. Actually, this is not the case. If this tube becomes weak, the phase of the oscillator signal will tend to advance. This change will be noted in the phase-detector circuit. One of the diode sections will conduct more heavily than the other and a positive dc correction voltage will be developed. This positive correction will cause the reactance tube to increase in conduction, and the phase of the oscillator signal will be retarded until it is correct. For this reason, the reactance tube itself should not be suspected of causing an error in the phase of the oscillator signal.

On the other hand, the tube or diodes used as phase detectors might well be suspected. If one of these becomes considerably weaker than the other, they will not conduct equally when the oscillator signal has the correct phase. As a result, a correction voltage will be applied to the reactance tube and will change the current through this tube. This will change the phase of the oscillator signal so that it will be incorrect. To a certain extent, this trouble may be compensated for by a readjustment of the hue control; however, if a considerable change from the normal setting of this control is needed, a definite trouble exists. Most late-model, tube-type color receivers employ semiconductor diodes in place of a vacuum tube for the phase detectors.

Defective Components

If the replacement of tubes or the adjustment of controls does not correct the phase of the oscillator signal, the chassis should be removed from the cabinet for further analysis. To make sure that the burst-amplifier stage is operating properly, the signal at the input of the phase detectors should be observed on the oscilloscope. The normal waveform of this signal should appear as shown in Fig. 11-11.

Under certain conditions, the entire chrominance signal instead of just the burst signal may appear at the inputs of the phase detectors. This would result if the bias on the burst amplifier were to go positive, in which case all of the chrominance signal at the grid would be amplified and applied to the phase detectors. A waveform that shows the signal at the input of one of the phase detectors

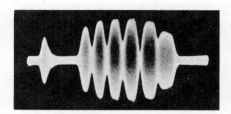

Fig. 11-20. Waveform at the phase detector when the chrominance signal is not keyed out by the burst amplifier.

when the entire chrominance signal is amplified can be seen in Fig. 11-20.

The phase of the chrominance signal is different for each color it represents. The phase detectors cannot follow rapid phase changes; consequently, the correction voltage applied to the reactance tube is developed from the average phase of the chrominance signal and is not correct. The oscillator signal will not have the proper phase, and the colors reproduced on the screen will be incorrect.

If the burst signal at the inputs of the phase detectors appears to be normal, the dc voltages at the cathode and at the anode of the phase detector should be measured with a vtvm. The voltage at the cathode will be positive, and the voltage at the anode will be negative. If one of these voltages is higher than normal and the other is lower than normal, resistor R1 or R2 may have increased in value or capacitor C1 or C2 may have changed in value.

If the voltages on the phase detectors appear to be normal, the dc correction voltage applied to the grid of the reactance tube should be measured with a vtvm or FET meter. When the receiver is tuned to a signal from a color-bar generator, this voltage should measure between a minus .5 and a plus .5 volt, depending on the setting of the hue control. One reason for an abnormal voltage at this grid could be a misadjustment of reactance coil L1. Another reason may be that the reactance tube is weak; however, as previously mentioned, a weak reactance tube will probably not cause the hues to be reproduced incorrectly.

The operation of some solid-state color-sync circuits is similar to the tube-type circuit just discussed. Fig. 11-21 shows the color-sync circuit used in a solid-state receiver. This circuit uses a varactor diode in place of the reactance tube employed in Fig. 11-19. The correction voltage developed by the chroma-sync phase detectors is applied to varactor diode X37. The operating frequency of the 3.58-MHz oscillator is determined by the 3.58-MHz crystal and the capacitance represented by C198, C199, and the varactor diode. Since the ca-

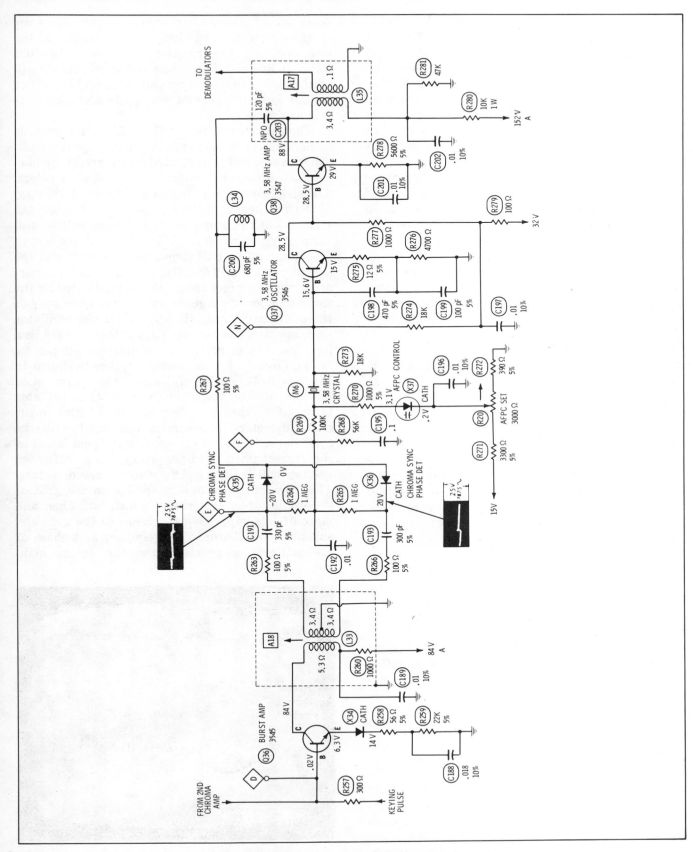

Fig. 11-21. Color-sync circuit used in a solid-state color receiver.

pacitance of the varactor diode is dependent on the applied voltage, the correction voltage from the chroma-sync phase detectors can be used to control the frequency and phase of the 3.58-MHz oscillator. The afpc set control, R20, is used to set the frequency of the oscillator. This circuit is subject to the phase-detector problems described for the tube-type color-sync circuit. Another source of phase error in this circuit is a defective varactor diode.

Some color-sync circuits do not have color phase detectors. Such a circuit is shown in Fig. 6-21 of Chapter 6. In this arrangement, the amplified burst signal is applied directly to the 3.58-MHz crystal. Consequently, the crystal output has the same frequency and phase as the burst signal. Ringing circuits of this type are found in both tube-type and solid-state color receivers. Phase errors in these circuits can be due to a component defect that causes a phase shift of the 3.58-MHz signal after the crystal. Such a defect is most likely to be in the hue control circuit. Another possible source of phase error is misalignment of the 3.58-MHz output transformer.

Misalignment

Wrong colors may also result if any of the tuned circuits in the color-sync section are not aligned properly. Normal operation can sometimes be restored by adjustment of the hue control; however, if this control has to be changed to a setting that is far from normal in order to obtain the proper colors, an alignment of the color-sync section may be in order.

If it is suspected that the color-sync section is not aligned properly, it should be checked, using the alignment procedure given in the service data. The color-sync alignment procedure for most color receivers is not complicated, and it can usually be performed with a vtvm or FET meter. If the colors are still wrong after the color-sync alignment has been completed, there is probably a defective component in the color-sync or demodulator circuit.

LOSS OF COLOR SYNCHRONIZATION

Whenever the chrominance and reference signals are arriving at the color demodulators, the chrominance signal will be demodulated and color will be reproduced, provided that the circuits that accomplish demodulation and those that handle the color signals after demodulation are operating properly. It has been shown in previous discus-

sions that if either the chrominance signal or the reference signals are absent at the inputs of the demodulators, the receiver will not reproduce color. It has also been shown that the colors will not be right unless the chrominance and reference signals have the correct amplitude and phase relationships.

When the frequency of the reference signals is incorrect, loss of color synchronization will result. This trouble will be indicated on the screen by horizontal or diagonal stripes of variegated colors. The stripes may be either in motion or stationary. If the frequency of the oscillator is unstable, the stripes will be in motion. The stripes will be stationary if the oscillator is stabilized at the wrong frequency. Fig. 11-22 shows the appearance of the screen when the 3.58-MHz oscillator is not operating at the correct frequency. As indicated by the small number of diagonal stripes, the frequency of the oscillator is only slightly off. If the oscillator were operating far off frequency, there would be a large number of horizontal or diagonal stripes.

The circuit of a color-sync section is shown in Fig. 11-19. Let us briefly review the operation of this circuit. The output of the 3.58-MHz oscillator is amplified; then a portion of it is fed back to the phase detectors where it is compared with the color-burst signal. If the oscillator signal is not of the correct phase and frequency, a dc correction voltage will be developed by the phase detectors. This correction voltage is applied to the grid of the oscillator-control stage, which will then add more or less capacitive reactance to the grid circuit of the oscillator. The frequency and phase of the oscillator are precisely controlled in this manner.

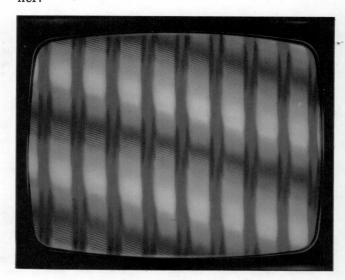

Fig. 11-22. Loss of color synchronization.

When the reference signal begins to lag behind the burst signal, the following events occur. Phase detector A conducts more than detector B. The negative dc correction voltage that is developed is added to the bias of the oscillator control stage. The gain of the control tube is reduced, and the amount of capacitance in the grid circuit of the oscillator is effectively decreased. The frequency and phase of the oscillator will tend to advance until the oscillator is in step with the color-burst signal. If the reference signal begins to lead the burst signal, a positive correction voltage is developed and opposite actions occur.

If anything should happen in the color-sync circuit to change the value of the dc correction voltage, the reference signal would be caused to change and would be thrown out of step with the incoming burst signal. When color synchronization is lost, the cause of the trouble will usually be located in some circuit ahead of the 3.58-MHz oscillator.

If the pattern on the screen appears like that shown in Fig. 11-22, and if the color-sync circuit under consideration is the type shown in Fig. 11-19, the first thing to check is the oscillator-control tube, V2. It would also be necessary to check the burst amplifier and the phase detectors, because color synchronization would be lost if one of these tubes were inoperative.

If replacement of tubes does not cure the trouble, a check of the circuit is in order. When troubleshooting the circuit, the best place to begin is in the oscillator-control stage. By making voltage measurements, the technician can determine whether or not the operating voltages in this stage are correct. With abnormal plate or screen voltages at the oscillator-control tube, the oscillator will drop out of synchronization.

The next step is to make sure the oscillator is able to operate at the correct frequency, with reasonably good stability, without being controlled by the phase detector. This can readily be checked by disabling the correction voltage fed to the oscillator-control tube, and then attempting to adjust the oscillator.

The method used to clamp the control tube grid at zero volts dc must not interfere with the rf operation of the stage. Thus, a grounding jumper may be connected to point C (or to some point in the burst amplifier, if so recommended in the service data), but not directly to the control tube grid. The color bars should "float" or move slowly across the screen, indicating that the oscillator will operate at the correct frequency.

If a check of voltages, tubes, and waveforms indicates no definite trouble, then the color-sync circuits should be aligned according to the service data.

When loss of color sync is encountered in a solid-state color receiver, the color-sync alignment should usually be checked first. In some solid-state receivers, the only color-sync adjustment is an afpc control. If alignment does not produce normal color sync, the burst-amplifier circuit should be checked next. If a normal burst signal is present at the color phase detectors, the phase-detector circuit may be at fault. An ohmmeter can be used to check for an open or shorted diode. In the circuit shown in Fig. 11-21, the loss of color sync may be due to a defective varactor diode. The color circuit shown in Fig. 11-3 uses an IC to process the burst signal and provide color sync. Loss of color sync in this type of circuit is normally due to a defective IC.

ALIGNING THE COLOR RECEIVER

The process of aligning the circuits in a color receiver is somewhat more involved than the alignment of the average monochrome receiver. The main reason for this is that the color circuits requiring alignment are not found in a monochrome receiver. In addition, the alignment of the rf and i-f circuits in a color receiver must be very precise in order to reproduce colors faithfully. For example, the video i-f section must have a bandwidth of nearly 4.5 MHz in order to pass the upper sidebands of the 3.58-MHz color subcarrier.

The sections that require alignment in a color receiver are the rf- i-f, sound, chroma-bandpass, and color-sync circuits. The rf, i-f, and sound circuits have their counterparts in monochrome receivers. Therefore, the alignment procedures for these circuits are very similar in both monochrome and color receivers. The chroma-bandpass and color-sync circuits are not used in a monochrome receiver. This discussion of color-receiver alignment will begin with those circuits that pertain to the color section. The circuits to be aligned will be presented in the following order: chroma-bandpass, color-sync, rf and i-f, and sound i-f. It should be noted that in order to align the chroma-bandpass circuits in most color receivers, the rf and i-f circuits must be correctly aligned first.

Aligning the Chroma-Bandpass Amplifier

The chroma-bandpass amplifier is the first stage in the chrominance channel of a color receiver. The

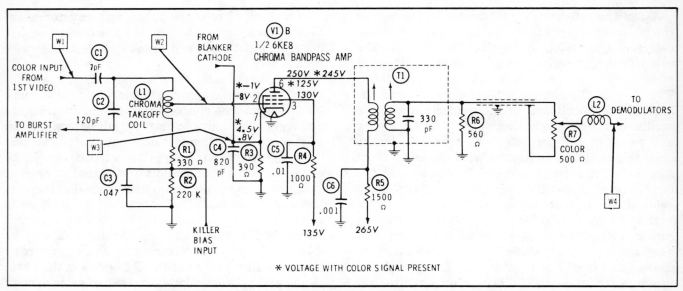

Fig. 11-23. Chroma-bandpass amplifier used in a tube-type color receiver.

signal at the bandpass-amplifier input consists of the chrominance and luminance signal, the sync and blanking pulses, and the color burst. At the output of the bandpass amplifier, only the chrominance signal remains. Therefore, the luminance signal must be blocked and the sync pulses, blanking pulses, and the color burst must be keyed out. Only the chrominance signal, which consists of the subcarrier sidebands that extend .5 MHz on each side of 3.58 MHz, are allowed to pass.

The bandpass-amplifier circuit shown in Fig. 11-23 is employed in a tube-type color receiver. The bandpass amplifier is similar to a conventional video amplifier, but it has components which are tunable so that the circuit can be aligned to provide the proper passband for the chrominance signal. When bandpass transformer T1 and chroma takeoff coil L1 are properly aligned, all of the sidebands of the color subcarrier will pass through the circuit to the color demodulators. Misalignment may cause a change in color saturation, loss of color detail, and poor color fit. Fig. 11-24 shows the normal response curve for the chroma-bandpass amplifier in Fig. 11-23.

If the bandpass-amplifier circuit is suspected of causing loss of color detail, the response of this circuit should be checked. The response curve for the bandpass amplifier is usually obtained by applying a chroma sweep signal to the grid of the bandpass amplifier stage and connecting an oscilloscope through a demodulator probe to the input of one of the color demodulators.

The video-detector load circuit affects the video i-f response curve, causing it to fall off in the re-

gion of the chrominance signal. The chroma take-off coil, L1, is adjusted to compensate for the slope of the response curve after the video detector. Therefore, when aligning the chroma-bandpass circuit, a method must be used that will take into account the loading effect of the video detector circuit. The alignment method used to accomplish this is called *video-frequency sweep modulation*, or *vsm*.

The vsm alignment procedure involves feeding a video-sweep modulated rf signal into the antenna terminals or a video-sweep modulated i-f signal into the mixer input. The method used depends on the type of alignment equipment available and the procedure given in the service data for the receiver being aligned. Many late-model sweep generators include built-in vsm alignment capability, which makes the chroma-bandpass alignment easier. When using this type of equipment to check

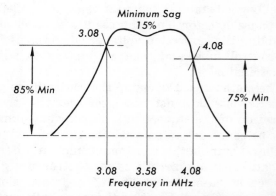

Fig. 11-24. Frequency-response curve for the chroma-bandpass amplifier shown in Fig. 11-23.

the overall chroma-bandpass alignment, refer to both the receiver service information and the operating manual for the sweep generator being used.

Aligning the Color-Sync Section

The function of the color-sync section is to supply the color demodulators with the 3.58-MHz cw signals that are used as reference signals to demodulate the chrominance signal. When two demodulators are used, two 3.58-MHz reference signals are needed; if all three primary colors are demodulated, three 3.58-MHz reference signals are required. These reference signals must have the correct phase relationship with each other and with the color burst so that the colors can be reproduced properly.

An oscillator circuit in the color-sync section generates a cw signal that is compared in frequency and phase with the color burst. This locally generated cw signal must have the same frequency as the color burst, and it must also have a certain phase relationship to the color burst. One of the purposes for alignment of the color-sync section is to ensure that the correct frequency and phase relationships are established. In addition, the cw reference signal must undergo a specific phase shift before it is applied to each of the color demodulators. Therefore, the second purpose for alignment of the color-sync section is to ensure that the exact amount of phase shift is obtained.

The reference subcarrier generated by the 3.58-MHz oscillator must have exactly the same frequency as the color subcarrier produced at the transmitter (3.579545 MHz). Furthermore, the original and regenerated subcarriers must agree in phase within ±5 degrees, so that the phase modulation of the sidebands can be accurately detected.

This high degree of precision is accomplished by using a crystal-controlled reference oscillator, stabilized by a closed-loop control system that is sensitive to minute amounts of drift. The synchronizing signal for the control system is a short burst of the original carrier that is transmitted during the horizontal blanking period. Only about 8 cycles of the subcarrier are included in each burst; the control system must keep the oscillator locked-in for approximately 220 cycles between the color bursts.

The hue (tint) control can often be used as a means of checking the operation of the color-sync section. When the correct color is obtained, the hue control should be approximately in the center

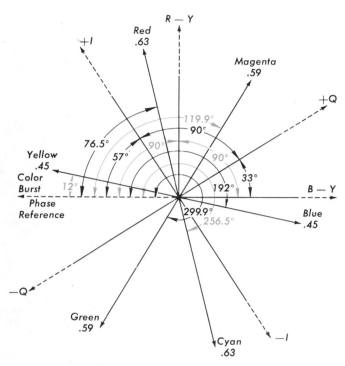

Fig. 11-25. Color-phase diagram.

of its range. This control should shift the phase of the color signals at least 30 degrees in either direction. The color-phase diagram shown in Fig. 11-25 is very useful for analysis of a color-bar pattern on the screen of a color receiver.

When a station signal is being received, normal flesh tones should be obtained with the tint control in the center of its range. The flesh tones should change from green to magenta as the tint control is rotated from one extreme to the other. When a

Fig. 11-26. Normal keyed-rainbow pattern.

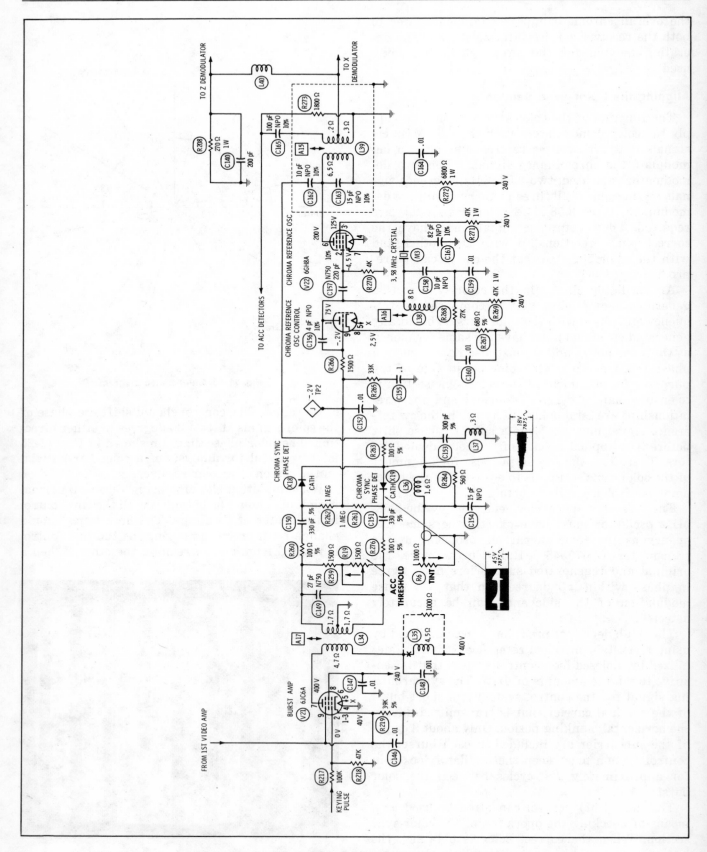

Fig. 11-27. Color-sync circuit used in a tube-type color receiver.

keyed-rainbow color-bar pattern is used, the green bar on the right side of the screen should reappear on the left side when the tint control is rotated from one end to the other. A normal keyed-rainbow pattern is shown in Fig. 11-26. If improper tint-control range or poor color synchronization is apparent, the color-sync alignment should be checked.

The color-sync section for a typical tube-type color receiver is shown in Fig. 11-27. The color-sync alignment procedure for this receiver is detailed in the following steps.

1. Connect a color-bar generator to the antenna terminals and adjust the receiver for a normal color-bar pattern. Set the color-killer control fully counterclockwise. Set the tint control to the center of its range.

2. Connect the control grid, pin 2, of the burst amplifier to ground. The prevents the burst signal from reaching the color phase detectors and allows the 3.58-MHz oscillator to operate at its free-running frequency. Connect the dc probe of a vtvm or FET meter through a 470K resistor to point K.

3. Adjust 3.58-MHz output transformer A15 for maximum indication on the meter. If no reading is obtained, the oscillator is probably not operating. Adjust oscillator control coil A16 to start the oscillator, and then adjust A15 for maximum indication on the meter. Remove the ground connection from pin 2 of the burst amplifier. Adjust burst transformer A17 for maximum indication on the meter. Make sure that the oscillator is running and locked in.

4. Short point J to ground. Remove the meter from point K. Adjust oscillator control coil A16 until the color bars stand still or drift slowly across the screen. Remove the short from test-point J and make sure that the color bars will sync with a low-level signal. If necessary, retouch A16 for best hold.

5. Connect the vertical input of an oscilloscope to the control grid of the red gun; connect the common lead to ground. Adjust the oscilloscope to obtain two stationary cycles of the waveform (7875 Hz). With the tint control in the center of its range, the waveform should look like the waveform pattern shown in Fig. 11-28. Check the range of the tint control. The bars should move 30 degrees either side of a normal pattern. If necessary, retouch A17 for proper range of the tint control. Also check the waveforms at the control grids of the blue and green guns. Fig. 11-29 shows a normal waveform at the blue gun, and Fig. 11-30 shows a normal waveform at the green gun.

Proper phasing of the demodulator outputs $(R - Y, B - Y,$ and $G - Y)$ can also be confirmed by viewing the screen with two of the crt control grids disabled. With a color-bar pattern on the screen, ground the blue and green crt control grids through individual 100K resistors. A correct presentation of the $R - Y$ signal under these conditions is shown in Fig. 11-31. The sixth color bar should be the same brilliance as the background (the space between the bars). This condition should be obtained with the tint control at approximately the center of its range. The tint control should have enough range to shift the pattern 30 degrees in either direction.

To check for the correct phase of the demodulated $B - Y$ signal, remove the 100K resistor from the blue crt control grid and connect it to the control grid of the red gun. This allows a normal $B - Y$ signal to reach the blue gun. The pattern that should be obtained is shown in Fig. 11-32. The third and ninth bar should have the same brilliance as the space between the bars.

Remove the 100K resistor from the green crt control grid and connect it to the blue crt grid. This allows a normal $G - Y$ signal to reach the green crt grid. The correct pattern for the demod-

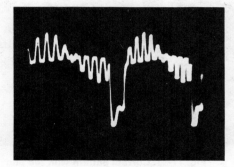

Fig. 11-28. Signal at the grid of the red gun.

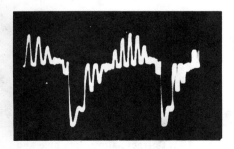

Fig. 11-29. Signal at the grid of the blue gun.

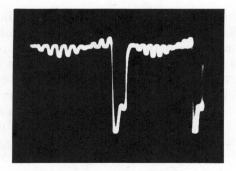

Fig. 11-30. Signal at the grid of the green gun.

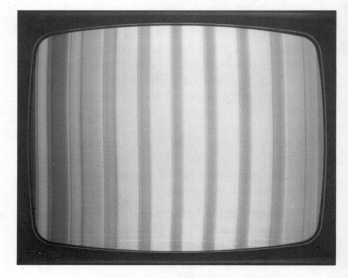

Fig. 11-32. Correct pattern with red and
green beams turned off.

ulated G − Y signal is shown in Fig. 11-33. The
seventh color bar should have the same brilliance
as the background. After removing the 100K resistor from the blue and red crt control grids, a
normal color-bar pattern should appear on the
screen.

The color-sync alignment procedures for other
circuits may vary considerably from the one just
given. This is especially true of solid-state color-
sync circuits. As a general rule, the color-sync
alignment procedure for solid-state receivers is
somewhat simpler than the procedure for tube-
type color-sync circuits. For example, color-sync
adjustment for many late-model Zenith color receivers consists of moving an alignment switch to
the ALIGN position and adjusting the APC control
for minimum movement of the color. The procedure for other solid-state color receivers may be
a little more involved. Unless you are completely
familiar with the procedure for a particular receiver, always refer to the service data when aligning a color-sync circuit.

Troubles in the burst-amplifier and color-sync
circuits have varying effects on the color picture.
A quick check of the operation of these circuits
can aid greatly in determining the origin of a defect and can save valuable time by concentrating
your troubleshooting effort on the defective stage.
Alignment of the color-sync section will usually
not be required unless a part has been changed or
unless the adjustments have been tampered with.
In some tube-type color-sync circuits, it is necessary to adjust the oscillator control coil after the
oscillator tube has been replaced.

Aligning the RF and I-F Sections

The functions of the rf and i-f sections in a color
receiver are essentially the same as they are in a

Fig. 11-31. Correct pattern with blue and
green beams turned off.

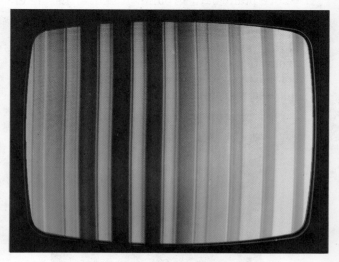

Fig. 11-33. Correct pattern with red and
blue beams turned off.

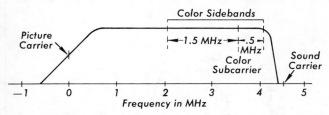

(A) Ideal response curve for the color receiver.

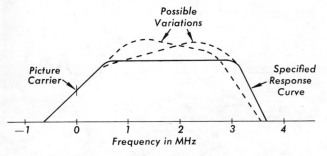

(B) Response curve for a monochrome receiver.

Fig. 11-34. Rf and i-f response curve for color and monochrome receivers.

quencies that are contained in the color picture signal and is considerably wider than the frequency range required in a monochrome receiver. (Refer to Fig. 11-34.)

An ideal rf and i-f response curve for a color television receiver is shown in Fig. 11-34A. This curve shows that the upper sidebands of the color subcarrier extend approximately 4.2 MHz above the video carrier. Consequently, a flat response up to 4.2 MHz is needed. In actual practice, the response curve may not be quite as flat as shown for the ideal curve. However, it should be as near to the ideal curve as possible so that all of the sideband frequencies will be amplified equally. Notice that the slope of the curve in the region of the sound carrier is very steep. It must be steep be-

monochrome receiver. These sections must provide amplification to a band of frequencies from .75 MHz below the picture-carrier frequency to about 4.2 MH above the picture-carrier frequency. This range of frequencies covers all of the sideband fre-

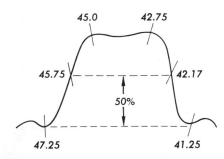

Fig. 11-36. Overall response curve for the video i-f section.

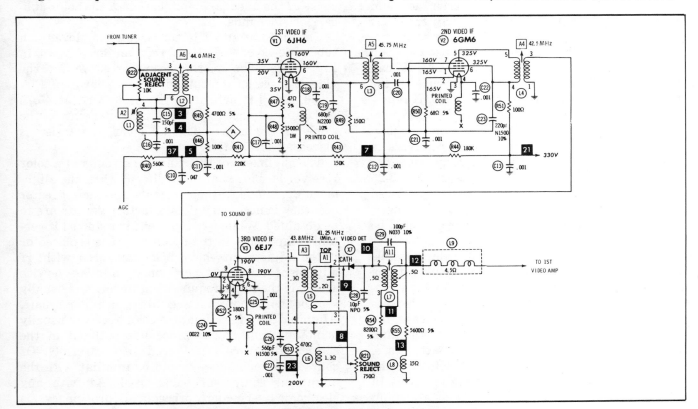

Fig. 11-35. Video i-f and detector circuits used in a tube-type color receiver.

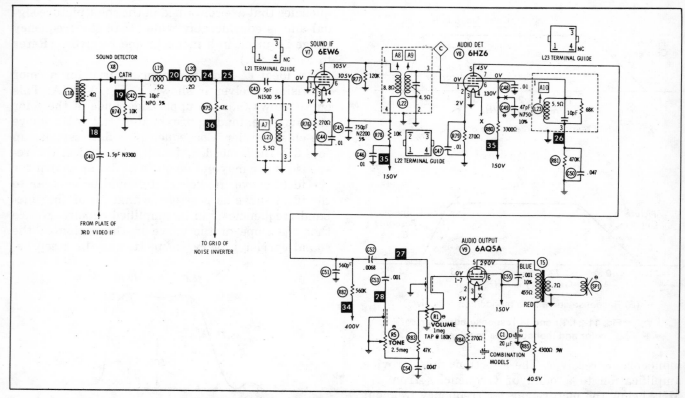

Fig. 11-37. Sound section of a tube-type color receiver.

cause the upper limit of the color-carrier sideband frequencies is less than 0.4 MHz away from the sound-carrier frequency.

By comparing the rf and i-f response curve in Fig. 11-34A to that shown in Fig. 11-34B, it can be seen that there is a notable difference in the bandpass. The curve shown in Fig. 11-34B represents the rf and i-f response curve for a typical monochrome receiver. The rf and i-f bandpass usually specified for most monochrome receivers is about 3.25 MHz. If the rf and i-f response curve of a color receiver should begin to taper off at 3.25 MHz, the color in a scene would not be reproduced because all of the upper, and part of the lower, sidebands of the chrominance subcarrier would be attenuated.

The dotted curves in Fig. 11-34B show possible variations in the rf and i-f response curve for a monochrome receiver. In one case, the high frequencies would be attenuated and the low frequencies would be overamplified. In the other case, the low frequencies would be attenuated and the high frequencies would be overamplified. In either case, such a response would have very little effect in the reproduced picture of a monochrome receiver. In a color receiver, however, the effect of a sloping response curve would be noticed more readily be-

cause of the unequal amplification of the color-subcarrier sidebands.

The video i-f and detector circuits employed in a tube-type color television receiver are shown in Fig. 11-35. The overall response curve for the video i-f section of this circuit is shown in Fig. 11-36. The i-f picture carrier is located at 45.75 MHz and is placed at the 50-percent point on the slope of the curve. The i-f color subcarrier is placed at 42.17 MHz.

The alignment of the rf and i-f circuits of a color receiver is somewhat more difficult than the alignment of a monochrome receiver because greater accuracy is necessary if satisfactory results are to be obtained. While i-f alignment is required if certain components in the i-f circuits are replaced or if the adjustments have been tampered with, rf alignment is rarely performed in most service shops. If the tuner requires alignment, it is usually sent out to a shop that specializes in tuner repair. Aligning the i-f section of a color receiver usually is not difficult if the procedure outlined in the service literature is followed very carefully. Of course, the technician should be familiar with the alignment equipment being used. As with any electronic servicing procedure, proficiency is gained through experience.

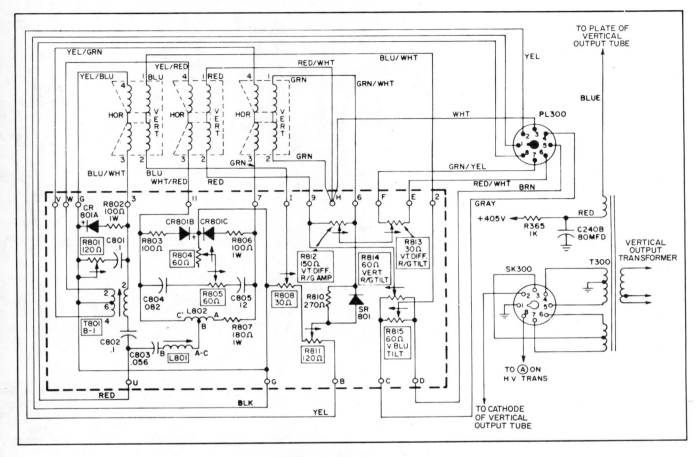

Fig. 11-38. Dynamic-convergence circuit.

Aligning the Sound I-F Section

A tube-type sound i-f section is shown in Fig. 11-37. This circuit uses a quadrature detector to demodulate the 4.5-MHz sound carrier. Before the quadrature detector can be aligned, the sound i-f transformers must be adjusted. Connect a vtvm through a detector probe to test-point C (the control grid of the 6HZ6 audio detector). Adjust the sound i-f input coil (A7) and the sound i-f transformer (A8 and A9) for maximum indication on the meter. Remove the vtvm and reduce the signal at the antenna terminals until background noise can be heard. Adjust the quadrature coil (A10) for maximum sound and minimum distortion.

The alignment procedure for solid-state sound i-f circuits or for tube-type circuits using a ratio detector will be somewhat different from the procedure just given. When aligning the sound i-f section, always consult the service data for the correct procedure. Many solid-state color receivers have fewer sound i-f adjustments than the tube-type sound section shown in Fig. 11-37.

CONVERGENCE TROUBLES

A schematic diagram of a dynamic-convergence circuit is shown in Fig. 11-38. This circuit supplies the signals to the dynamic-convergence coils on the neck of the picture tube.

Whenever it is noticed that the beams are not converging properly on the screen, the adjustments of the convergence controls should be the first things checked. The setup procedure for obtaining beam convergence was discussed in Chapter 9.

If the screen shows that misconvergence is present in the vertical direction only, the controls for vertical convergence are adjusted first to see if the trouble can be corrected. When making convergence adjustments, remember that a signal from a pattern generator must be used.

If adjustment of the controls for vertical convergence does not cure the trouble, use the oscilloscope to make sure that the pulses from the vertical output transformer are being applied to the input of the circuit.

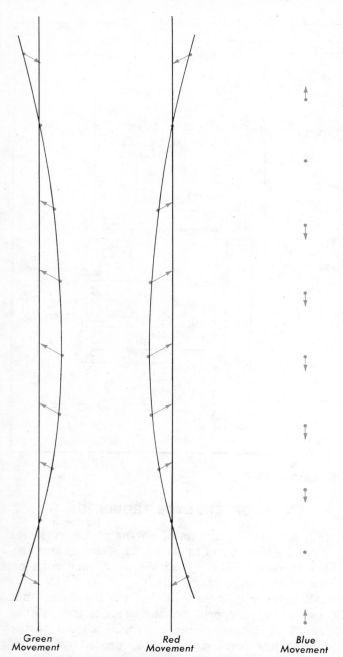

Fig. 11-39. Dot movements caused by adjustment of the vertical amplitude controls.

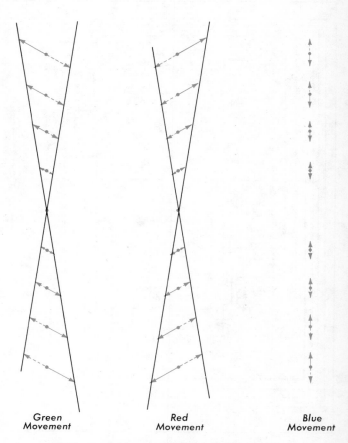

Fig. 11-40. Direction of dot movement caused by adjustment of vertical-tilt controls.

In the circuit of Fig. 11-38, each convergence coil has a control for vertical amplitude and one for vertical tilt. By adjusting each control while observing a dot pattern on the screen, the technician can determine which portion of the circuit is not functioning properly. Fig. 11-39 shows the directions in which the vertical center rows of dots should move when the amplitude controls are adjusted. Fig. 11-40 shows the directions in which the dots should move when the tilt controls are ad-

justed. When it has been determined which control is not functioning properly, a voltage and resistance check of the circuit associated with that particular control should reveal the cause of the trouble.

If the beams are misconverged in the horizontal direction but are properly converged in the vertical direction, only the horizontal convergence circuit need be checked. The same troubleshooting procedure as used for the vertical convergence can be followed. First, make sure that the horizontal pulse from the horizontal-output transformer is being applied to the convergence circuit. Associated with each convergence coil is a control for horizontal amplitude and one for horizontal phase. The directions in which the dots should move when the controls for horizontal amplitude are adjusted are shown in Fig. 11-41. If the dots do not move correctly when a control is adjusted, a voltage and resistance check of the circuit associated with that particular control should reveal the cause of the trouble.

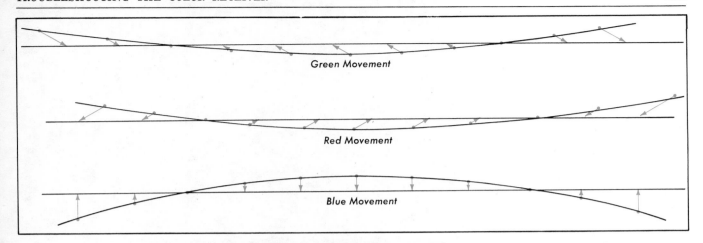

Fig. 11-41. Dot movements caused by adjustment of horizontal-amplitude controls.

QUESTIONS

1. What is the first step in troubleshooting a color receiver?

2. If a video signal from a color-bar generator will not produce color on the screen, in what sections of the receiver will the trouble be located?

3. What is the effect on the screen of a color receiver when the burst amplifier fails?

4. If a color-bar generator connected to the antenna terminals produces a normal color-bar pattern but the colors are improper when connected to the antenna, where will the trouble be located?

5. What effect will be noticed on the screen if the bandpass amplifier is weak?

6. If the hue of all colors is wrong, what is the trouble?

7. Can the trouble in Question 6 be caused by a defective tube? If the answer is yes, what tube functions will cause this trouble?

8. What service procedure should be used if color impurity is noticeable on the screen?

9. What is the effect on the screen of a color receiver when the reference oscillator is stabilized at the wrong frequency?

10. Can the video i-f section of a color receiver be aligned to the same bandpass as a monochrome receiver? Give the reasoning behind your answer.

Fig. 11-13. Dot movement caused by admixture of horizontal-amplitude controls.

QUESTIONS

Section IV

Appendices

Appendix A

Equations

From Chapter 1

These are the conversion equations used in the study of colorimetry to derive the three-dimensional co-ordinate values for x, y, and z:

$$x = \frac{x}{x + y + z} \qquad (1)$$

$$y = \frac{y}{x + y + z} \qquad (2)$$

$$z = \frac{z}{x + y + z} \qquad (3)$$

From Chapter 2

Line frequency,

$$f_L = \frac{4.5 \times 10^6}{286} = 15{,}734.264 \text{ Hz.}$$

Frame frequency,

$$f_F = \frac{f_L}{525} \times 2 = 59.94 \text{ Hz.}$$

Subcarrier frequency,

$$f_S = \frac{455}{2} \times f_L = 3.579545 \text{ MHz.}$$

From Chapter 3

The equations from this chapter and their identifications and numbers are as follows:

$$E_Y = .30E_R + .59E_G + .11E_B. \qquad (4)$$

Color-difference signals for red, green, and blue, respectively,

$$E_R - E_Y = .70E_R - .59E_G - .11E_B. \qquad (5)$$
$$E_G - E_Y = .41E_G - .30E_R - .11E_B. \qquad (6)$$
$$E_B - E_Y = .89E_B - .59E_G - .30E_R. \qquad (7)$$

In-phase color signal in terms of $E_R - E_Y$ and $E_B - E_Y$,

$$E_I = .74(E_R - E_Y) - .27(E_B - E_Y). \qquad (8)$$

Quadrature-phase color signal in terms of $E_R - E_Y$ and $E_B - E_Y$,

$$E_Q = .48(E_R - E_Y) + .41(E_B - E_Y). \qquad (9)$$

In-phase color signal in terms of the three primaries,

$$E_I = .60E_R - .28E_G - .32E_B. \qquad (10)$$

Quadrature-phase color signal in terms of the three primaries,

$$E_Q = .21E_R - .52E_G + .31E_B. \qquad (11)$$

Color picture signal,

$$E_M = E_Y + [E_Q \sin(\omega t + 33°) + E_I \cos(\omega t + 33°)]. \qquad (12)$$

Color-difference signal for green in terms of $E_R - E_Y$ and $E_B - E_Y$,

$$E_G - E_Y = -.51(E_R - E_Y) - .19(E_B - E_Y).$$

From Chapter 7

Color-difference signal for blue in terms of $R - Y$ and $G - Y$,

$$B - Y = -2.73(R - Y) - 5.36(G - Y).$$

From Chapter 8

These are the equations for the red, blue, and green signals, respectively, which are applied to the picture-tube guns in an I-Q receiver:

$$E_R = .96E_I + .63E_Q + 1.00E_Y. \qquad (13)$$
$$E_B = -1.11E_I + 1.72E_Q + 1.00E_Y. \qquad (14)$$
$$E_G = -.28E_I - .64E_Q + 1.00E_Y. \qquad (15)$$

Color-difference signals in terms of I and Q are rewritten from equations 13, 14, and 15, respectively, as follows:

$$E_R - E_Y = .96E_I + .63E_Q.$$
$$E_B - E_Y = -1.11E_I + 1.72E_Q.$$
$$E_G - E_Y = -.28E_I - .64E_Q.$$

Appendix B

The .877 and .493 Reduction Factors of the Color Signal

During the transmission of a color signal, the relative level of the modulation is held within certain limits to prevent overmodulation of the signal. If this were not done, severe overmodulation would occur during the transmission of fully saturated colors. If overmodulation were allowed to occur, the receivers would be overdriven and would therefore cause distortion in the reproduced picture and a possible buzzing sound in the audio.

The amplitude of the chrominance signal is effectively lowered by reduction of the amplitudes of the color-difference signals $E_R - E_Y$ and $E_B - E_Y$ which are contained in the two signals, I and Q, that make up the transmitted chrominance signal. The $E_R - E_Y$ signal is reduced by a factor of .877 and the $E_B - E_Y$ is reduced by a factor of .493. These reduction factors are shown on the vector drawing of Fig. B-1.

The amplitude of the luminance signal is not changed. This signal must be transmitted as specified by the equation,

$$E_Y = .30E_R + .59E_G + .11E_B, \qquad (4)$$

so that it will be compatible with monochrome standards.

DISADVANTAGES IF COLOR SIGNAL IS NOT REDUCED

Before discussing how the reduction factors were obtained, let us see what would happen if the chrominance signal were not reduced in amplitude. The formulation of the color picture signal when the fully saturated primary and secondary colors of a color-bar pattern are being transmitted is shown in Fig. B-2. In this case, the two color-difference signals $E_R - E_Y$ and $E_B - E_Y$ are combined to form the chrominance signal. The amplitude of the chrominance signal for each color bar shown is governed by the following equations for the color-difference signals:

$$E_R - E_Y = .70E_R - .59E_G - .11E_B, \qquad (5)$$
$$E_B - E_Y = -.30E_R - .59E_G + .89E_B \qquad (7)$$

The levels of the luminance signal are shown in Fig. B-2A. To the luminance signal would be added the chrominance signal which is shown in Fig. B-2B. This illustration shows the levels of the chrominance signal after the two color-difference signals have been combined.

Fig. B-2C shows the color picture signal after the signal in Fig. B-2B has been superimposed upon the luminance signal. Notice that overmodulation of the signal occurs in both directions. The signals for yellow, cyan, and green greatly overmodulate into the whiter-than-white region (beyond the white level); and the signals for magenta, red, and blue greatly overmodulate into the blacker-than-black region (beyond the black level). The signals for the colors yellow and blue are the worst offenders.

Fig. B-3 shows the amplitudes and the phases of the color signals when the color-difference signals have not been reduced in amplitude. When compared with a normal color-phase diagram (like the one which appears later in this appendix), it can be seen that the resultant color vectors in Fig. B-3 are much longer.

DETERMINING REDUCTION FACTORS

As a result of experiments that were performed early in the development of the transmitting standards for color television, it was found that over-

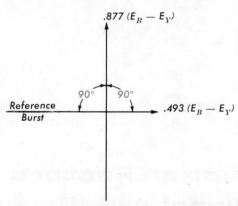

Fig. B-1. Vector diagram showing the color-difference signals and the reduction factors.

modulation of 33⅓ percent in either direction could be tolerated. This meant that the color signal could not go beyond the white level or the black level in excess of 33⅓ percent. In order to ensure this, the reduction factors of .877 and .493 for the color-difference signals were determined. The way in which these reduction factors were determined is as follows:

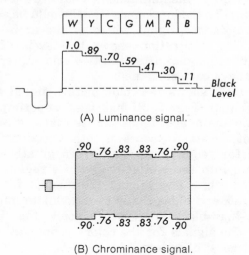

(A) Luminance signal.

(B) Chrominance signal.

(C) Color-picture signal.

Fig. B-2. Formulation of the color picture signal without the reduction of the chrominance signal. (This is a hypothetical case and the signals are not transmitted at the amplitudes shown.)

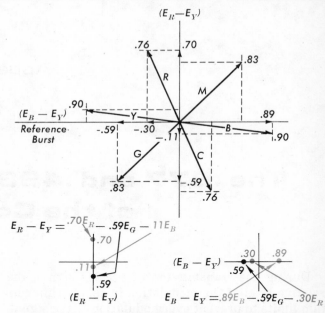

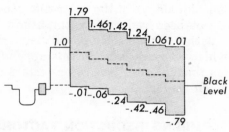

Fig. B-3. Vector diagram showing the amplitudes and phases of the color signals when the color-difference signals have not been reduced. (This is a hypothetical case and the vectors do not reach the actual magnitudes shown.)

The maximum permissible amplitudes of the subcarrier for the primary colors are first determined by inspection of the luminance signal. When red is being televised, the amplitude of the luminance signal is .30. A chrominance signal with an amplitude of .30 can be added to this luminance level without overmodulation being produced. Since an additional 33⅓ percent can be allowed, the total permissible amplitude of the red chrominance signal would be .30 plus .333, or .633. The maximum permissible amplitude for the blue chrominance signal can be determined in the same manner. The luminance signal for blue has an amplitude of .11; therefore, the allowable amplitude of the blue chrominance signal would be .11 plus .333, or .443.

Since red and blue are the two variables of the subcarrier signal, the limits of .633 and .443 for these two primaries can be used in determining the reduction factors.

Let us change the two expressions for the color-difference signals as follows:

$$E_R - E_Y = a \, (.70E_R - .59E_G - .11E_B),$$
$$E_B - E_Y = b \, (-.30E_R - .59E_G + .89E_B),$$

where,

  a and b are two reduction factors to be determined.

From this, it can be seen that the red chrominance signal would have a value of $.70a$ for the $E_R - E_Y$ component, and a value of $-.30b$ for the

$E_B - E_Y$ component. The blue chrominance signal would have a value of $-.11a$ for the $E_R - E_Y$ component, and a value of $.89b$ for the $E_B - E_Y$ component. From this data, the two matrix equations can be set up as follows:

$$R = \sqrt{(.70a)^2 + (-.30b)^2} = .633,$$
$$B = \sqrt{(-.11a)^2 + (.89b)^2} = .443.$$

The values of a and b can be found by solving these two expressions as a set of two simultaneous equations.[1] In so doing, it would be found that a equals .877 and b equals .493, which are the reduction factors for the $E_R - E_Y$ and $E_B - E_Y$ signals.

AMPLITUDE OF COLOR SIGNAL AFTER REDUCTION

The waveforms shown in Fig. 8-4 result when the color-bar pattern is scanned and when the amplitudes of the color-difference signals have been reduced by the factors of .877 and .493. Fig. B-4A is the luminance signal, and Fig. B-4B shows the envelope of the chrominance signal. Notice that the overall amplitude is somewhat lower than that which was shown in Fig. B-2B. For instance, the chrominance signal for the color red has a maxi-

---

1. In order to arrive at the exact values of the reduction factors (.877 and .493), the precise values of the constants in the above equations must be used as follows:
Let us solve for a^2 in the following expression:

$$\sqrt{(.701a)^2 + (.299b)^2} = .299 + .333.$$

Squaring both sides of the foregoing expression, we have:

$$(.701a)^2 + (.299b)^2 = (.632)^2$$
$$.4914a^2 + .0894b^2 = .3994.$$

Transposing $.0894b^2$ and dividing both sides by .4914, we find:

$$a^2 = .8127 - .1819b^2.$$

Let us solve for a^2 in the following expression:

$$\sqrt{(.115a)^2 + (.885b)^2} = .115 + .333.$$

Squaring both sides of the foregoing equation, we have:

$$(.115a)^2 + (.885b)^2 = (.448)^2$$
$$.0132a^2 + .7832b^2 = .2007.$$

Transposing $.7832b^2$ and dividing both sides by .0132, we find:

$$a^2 = 15.2045 - 59.3333b^2.$$

We now have two expressions for a^2; therefore:

$$.8127 - .1819b^2 = 15.2045 - 59.3333b^2.$$

Transposing, we have:

$$59.1514b^2 = 14.3918.$$

Dividing both sides by 59.1514, we have:

$$b^2 = .2433$$
$$b = .493.$$

Substituting the value for b^2 in the expression

$$a^2 = .8127 - .1819b^2,$$

we have:

$$a^2 = .8127 - .1819(.2433)$$
$$a^2 = .7684$$
$$a = .877.$$

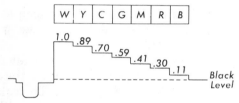

(A) Luminance signal.

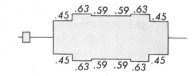

(B) Chrominance signal.

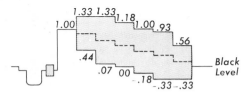

(C) Color-picture signal.

Fig. B-4. Formulation of the color-picture signal when the amplitudes of the color-difference signals have been reduced.

mum amplitude of .63 as compared to .76 before being reduced. The $E_R - E_Y$ component of red has an amplitude of .61 after reduction, and the $E_B - E_Y$ component has an amplitude of $-.15$ after reduction.

Fig. B-4C shows the color picture signal after the chrominance signal has been superimposed

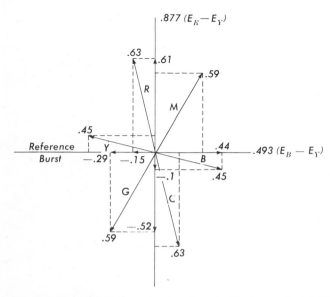

Fig. B-5. Vector diagram showing the amplitudes and phases of the color signals after the color-difference signals have been reduced.

upon the luminance signal. It can be seen that any overmodulation has been held within the specified limit of 33⅓ percent.

A signal of this magnitude can be handled without any difficulty occurring either at the transmitting end or at the receiving end. The color signal very seldom reaches the amplitudes shown in Fig. B-4C because these amplitudes represent the values that will be attained only when fully saturated colors at maximum luminance are being tele-vised. This very rarely happens under normal televising conditions because colors of this type almost never appear in normal subject material. About the only way they can appear is by artificial generation such as by a color-bar generator.

The color-phase diagram in Fig. B-5 shows the amplitudes and phases of the color vectors after the amplitudes of the color-difference signals have been reduced.

Appendix C

A Study of Vectors

It is one thing to be able to recognize a vector diagram; it is quite another to be able to understand all that such a diagram means in terms of quantities and operations that are familiar. Until recently, vectors have played a minor role in radio and tv servicing principally because the phase relationships between signals have been of little importance. Color television has changed this situation. Phase must be designated in order to specify accurately a color-picture signal, and vector diagrams are used for this purpose. The service technician should be able to recognize a vector diagram when he sees one, and he should be able to interpret its meaning with the same ease as he would an equation in Ohm's law.

Plane geometry and trigonometry form the mathematical basis for any study of vectors; therefore, the reader is advised to brush up on these subjects as a preliminary step. In this discussion, however, an attempt has been made to get by with as little reference to geometry and trigonometry as possible. Only the most basic fundamentals of these subjects are used.

Fig. C-1 shows a diagram which is typical of those encountered in color-television work. It is called a color-phase diagram and depicts some of the various signals which are frequently present during tests of color receivers. The purpose of this discussion is to provide the reader with helpful information about vector diagrams such as the color-phase diagram shown in Fig. C-1.

DEFINITION

A vector is the symbol which denotes a directed quantity; that is, it expresses a quantity which can only be completely described in terms of a magnitude and a direction. Wind velocities, voltages and current of electricity, and forces of all kinds are directed quantities and therefore can be expressed as vectors. Some quantities do not have the added property of direction, and these are called scalars to distinguish them from vectors. Examples of scalars are: the volume in a container, the resistance in a wire, and the inductance in a coil.

Since the vector symbol must show magnitude and direction, it takes the form of a straight line having a specific length and terminated at one end with an arrowhead. The length of the line represents the magnitude, and the arrowhead indicates the direction of the vector quantity. The specific points at which a vector begins and terminates have no significance; this means that a vector can be equivalently expressed by any line drawn parallel to it and of equal length. The latter characteristic of vectors is an important one and should be kept in mind at all times. It will be demonstrated pointedly in portions of the following discussion.

TRACTOR-AND-LOAD ANALOGY

As mentioned previously, a force is a directed quantity and can be denoted by a vector. The force with which we are primarily concerned in the electronics industry is voltage, sometimes referred to as electromotive force (emf). Because the force that is voltage is somewhat less tangible than certain other types of forces, it might be well to preface an explanation of voltage vectors by an analogy that is based on a more obvious kind of force.

Let us assume that we have a load to which we can connect a tractor. If the tractor is started and the connecting rope or chain is drawn taut, a force will be exerted upon the load. The nature of this force will be dependent upon the pulling power of the tractor and upon the direction in which the tractor is headed. Fig. C-2A shows the tractor

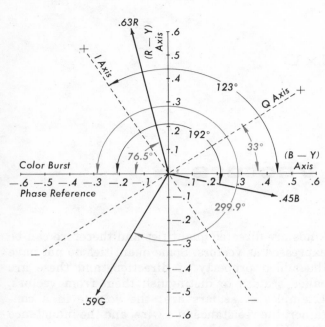

Fig. C-1. Color-phase diagram.

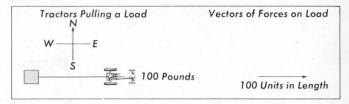

(A) Force of 100 pounds toward the east.

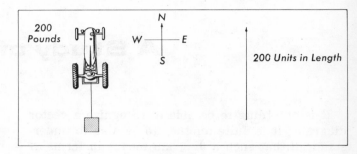

(B) Force of 200 pounds toward the north.

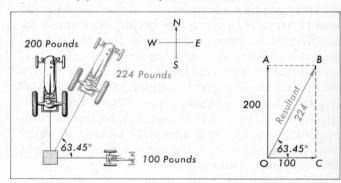

(C) Forces of 100 pounds east and 200 pounds north.

Fig. C-2. Forces and their vectors.

pulling toward the east with a force of 100 pounds. The vector which describes this force is drawn in the right column of the illustration. Note that it points toward the right and is 100 units in length. With one unit of length equivalent to a pound of force, the vector can be said to provide a complete description of the force which is exerted on the load by the tractor.

A limiting factor should be introduced at this point. In order to avoid certain misconceptions which would otherwise become obstacles during later stages in the development of this analogy, let us assume in all of the tractor-and-load problems that the load is too great to be moved by the forces exerted upon it. This limitation can be made since forces exist regardless of whether or not they cause movement. A man leaning against a brick wall may not cause the wall to move, but he still exerts a force upon it. Similarly, even though the tractor in these problems does not move its load, it still exerts a force upon the load.

In Fig. C-2B, the tractor is shown headed in a northerly direction and exerting a force of 200 pounds on the load. The vector which describes this force is drawn in the right column of the illustration. It points upward and has a length of 200 units.

Two or more vectors can be added to each other just as numbers can be added. The sum in vector addition is called the resultant vector. Fig. C-2C shows two tractors attached to a single load but pulling in different directions and with different amounts of force. The problem is to replace the two tractors with one tractor and to keep the total force on the load exactly the same. If vector addition is performed, the resultant vector will specify the magnitude and direction of the force to be provided by the substitute tractor. Vectors OA and OC in the right column of Fig. C-2C denote the forces exerted by the two original tractors. These vectors are drawn at right angles to each other and with lengths of 200 and 100 units. The resultant vector is found graphically by completing the parallelogram OABC. Dotted line AB is drawn parallel to vector OC, and dotted line BC is drawn parallel to vector OA. The resultant is the diagonal line OB. Measurement will show its length to be 224 units and its angle with vector OC to be 63.45 degrees. We can say, therefore, that the substitute tractor would have to pull with a force of 224 pounds and in a direction 63.45 degrees north of east in order to satisfy the requirements of the problem.

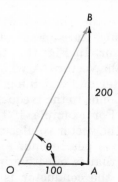

Fig. C-3. A second method of adding the vectors of forces in Fig. C-2.

A trigonometric rather than a graphical solution of the problem in Fig. C-2C may be performed. For this method of solution, the vector triangle in Fig. C-3 is used. Vectors OA and AB represent the forces exerted by the two tractors. Remember that vector AB can assume the position which it has, because a vector specifies magnitude and direction only—not point of action. Triangle OAB is a right triangle; therefore, by trigonometry:

$$\tan \theta = \frac{200}{100}.$$

From a table of trigonometric functions,

$$\theta = 63.45°.$$

To find the length of the resultant vector OB, use the expression:

$$\sin 63.45° = \frac{200}{\text{vector OB}}$$

Solving, we find:

$$\text{vector OB} = \frac{200}{\sin 63.45°} = \frac{200}{.894}$$
$$= 224.$$

If desired, the reverse of the foregoing operation can be performed. In other words, a vector may be broken into two or more component vectors which, when added together, produce the original vector. In Fig. C-4, the vector OA is given with a length of 300 units. Let us break this vector into rectangular components along the axes specified in the drawing. Triangle OBA is a right triangle; therefore,

$$\sin 47° = \frac{\text{vector BA}}{300}.$$

Solving, we find:

$$\text{vector BA} = 300 \sin 47° = 300 \times .731 = 219.3.$$

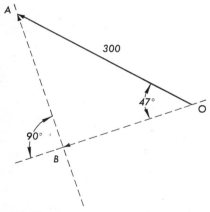

Fig. C-4. The rectangular components of a vector.

Furthermore, by using the relation:

$$\cos 47° = \frac{\text{vector OB}}{300},$$

we can find the other rectangular component:

$$\text{vector OB} = 300 \cos 47° = 300 \times .682 = 204.6.$$

In a similar manner, any number of pairs of rectangular components can be found for the vector OA in Fig. C-4, depending upon the axes selected. Axes that are horizontal and vertical are very commonly used.

ROTATING VECTORS

Up to this point, we have been discussing vectors which have remained static or unchanged with the passage of time. As a matter of fact, however, the vectors which depict many electrical quantities are dynamic; that is, they vary regularly with time. One such vector is the rotating vector which constantly changes its direction by revolving about its point of origin. An ac voltage is an example of a quantity that is best described by a rotating vector.

As a start toward understanding the rotating vector, an enlargement of the tractor-and-load analogy may be helpful. In Fig. C-5, the tractor is shown on a large turntable which revolves counter-clockwise. The load is on wheels which rest on a straight pair of tracks. The magnitude of the pulling force exerted by the tractor is assumed to be fixed, and the turntable rotates at a constant speed. In order to avoid certain problems which are incidental to this analogy, the load must be considered immovable in any direction despite the facts that it rests on a railway and that it has a force exerted upon it. The immobility of the load can be tolerated, since we are primarily interested

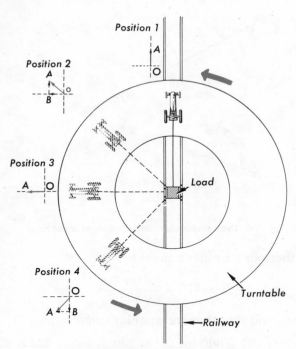

Fig. C-5. Tractor and load arranged to illustrate a rotating vector.

in that component force which tries at least to move the load along the railway.

When the tractor is in position 1, the total force exerted on the load is in the direction of the railway. This is indicated by vector OA in the vector diagram for position 1. When the tractor is in position 2, the vector diagram for that position shows that the force which is in line with the railway is the rectangular component BA. In position 3, the total force is exerted at right angles to the railway; therefore, there is no component in the direction of the railway. The last position that is shown on the drawing is position 4. The vector diagram for this position indicates that the rectangular component OB lies in the direction of the railway.

Let us replace the railway with an electrical conductor and substitute an ac voltage in place of the pulling tractor. In Fig. C-6, the ac voltage is represented by the vector OA which rotates about point O as a center and which has a speed of rotation that corresponds to the frequency of the alternating voltage. For example, if the frequency is one megahertz, the vector rotates at a speed of one million revolutions per second. As vector OA rotates, the force that is effective in trying to produce current in the conductor is represented by the rectangular component BA which lies in the direction of the conductor. This component changes in magnitude and reverses direction as vector OA rotates. If the variation in component vector BA is plotted graphically with respect to time, the sine wave on the right side of Fig. C-6 is formed. The latter, of course, can be recognized as a very conventional way of illustrating an ac voltage. Once every revolution or cycle, the vector OA assumes the position that is shown for it in the vector diagram. At this instant in every cycle, the voltage in the conductor has a magnitude equal to the length of component vector BA and is conventionally assigned a positive polarity. It may be noted that whenever the ac voltage vector OA assumes a position in line with the conductor, a peak of voltage occurs in the conductor. On the other hand, whenever the ac voltage vector OA lies perpendicular to the conductor, there is zero voltage in the conductor.

Suppose that there are both ac and dc voltages in the conductor. Fig. C-7 shows the vector and sine-wave representations under this condition. The vector OA denotes the dc voltage; this vector does not rotate since its direction does not vary with time. The vector AB depicts the ac voltage and rotates about point A. Vector OB is the result of adding vectors OA and AB, and the rectangular

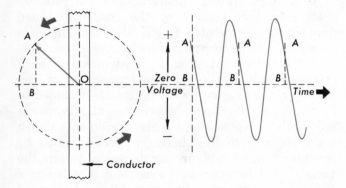

Fig. C-6. An ac voltage represented by a rotating vector.

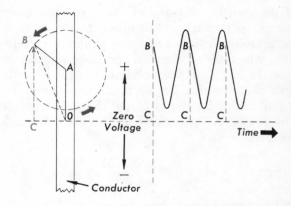

Fig. C-7. Dc and ac voltages represented by fixed and rotating vectors respectively.

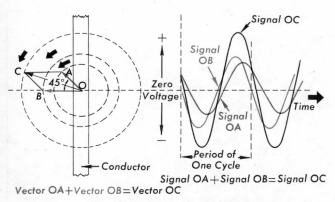

Vector OA+Vector OB=Vector OC

Fig. C-8. Two ac voltages as rotating vectors.

components of vector OB are given as vectors OC and CB. Of primary interest, as far as the voltage in the conductor is concerned, is the rectangular component CB. If plotted against time, component vector CB produces the sine wave at the right in Fig. C-7. The point of minimum positive voltage in each cycle is reached when the rotating vector AB directly opposes the dc voltage vector OA. The point of maximum positive voltage in each cycle is reached when the rotating vector AB is in line with and aids the dc voltage vector OA.

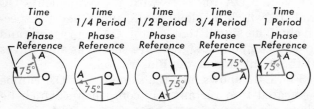

Fig. C-9. Vector and phase reference at five times during rotation.

Under certain conditions, there may be two ac voltages present in a conductor. This situation is illustrated in Fig. C-8. Vectors OA and OB represent the two ac voltages which, in this case, have the same frequency although they are separated in phase.

PHASE

The matter of phase should be discussed before going further in this study. Phase becomes important in this case of rotating vectors only; fixed vectors such as those used to denote dc voltages have no phase. Strictly speaking, phase is the position of a rotating vector during its period of rotation and is expressed in relation to the starting position or to some assumed standard position. It has become customary, however, to use the term phase to mean the phase difference between two vectors which are both rotating. In Fig. C-9, two vectors are shown at five different times during one period of rotation. Both vectors are moving at the same speed. The one labeled phase reference may or may not have a specific magnitude (note the absence of an arrowhead) because its most important property is that of direction. It is used so that the phase of other vectors in a diagram can be stated with respect to it. In Fig. C-9, for example, the vector OA can be said to have a lagging phase of 75 degrees. Note that this angle is preserved throughout rotation. In nearly all vector diagrams, the rotating vectors are pictured at an instant when the phase reference assumes a horizontal position. Thus, the usual vector diagram shows

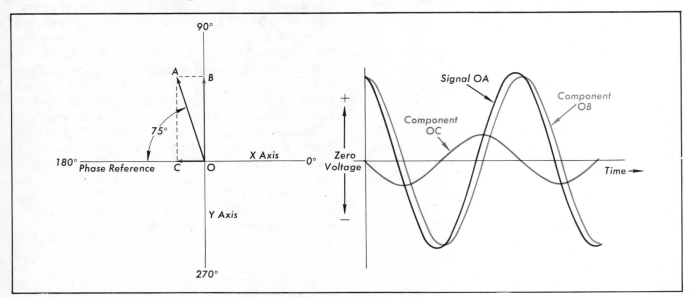

Fig. C-10. Rotating vector and its rectangular components along axes X and Y.

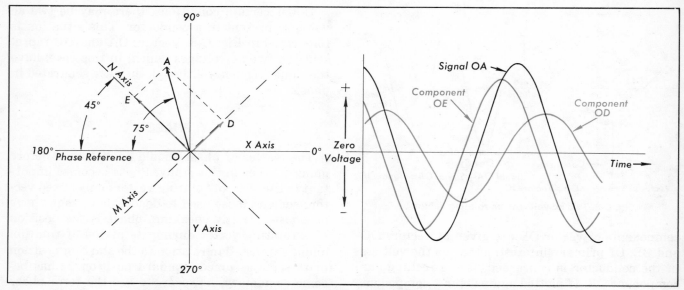

Fig. C-11. Rotating vector and its rectangular components along axes M and N.

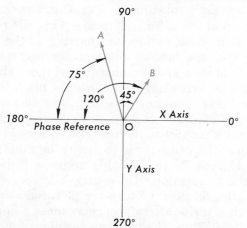

Fig. C-12. Vector diagram which describes signal conditions in a circuit at different times.

rotating vectors in positions corresponding to those at zero time and at half-period time in Fig. C-9. No matter which position may be pictured, the diagram still shows that vector OA lags the phase reference by 75 degrees.

Refer again to Fig. C-8. If vector OB is considered as the phase reference in this vector diagram, then vector OA can be said to have a lagging phase of 45 degrees. Vector OB, of course, is at zero phase. Vector OC is the resultant of adding vectors OA and OB. All three rotate at the same speed since they represent signals having the same

frequency. The signals differ in magnitude and in phase.

A rotating vector can be broken into rectangular components both of which will rotate at the same speed as the original vector. Fig. C-10 shows the vector OA together with its components OC and OB which lie along the X and Y axes, respectively. The signal that vector OA represents and its two component signals are shown as sine waves at the right side of the illustration.

Fig. C-11 shows how the same vector OA can be separated into rectangular components which lie along a different set of axes. The M and N axes have their positions specified by the 45-degree angle between the N axis and the phase reference. Vectors OD and OE are rectangular components of vector OA and lie on these axes. A sine-wave illustration of the vectors is included in the figure.

Sometimes one vector diagram is used to show the nature of two or more signals that appear singly at different times in a circuit. The color-phase diagram in Fig. C-1 is this type of diagram. A simpler example is given in Fig. C-12 which shows the nature of two signals represented by vectors OA and OB. The mere fact that both signals appear on the same vector diagram means that they have the same frequency. Their magnitudes are disclosed by their lengths, and their phases are given by the angles shown.

Appendix D

Glossary

angstrom units—A minute unit of length equal to one ten-thousandth of a micron, one-tenth of a millimicron, or one hundred-millionth of a centimeter (10^{-8} cm).

aperture mask—A thin sheet of perforated material which is placed directly behind the viewing screen in a three-gun color picture tube to prevent the excitation of any one color phosphor by either of the two electron beams not associated with that color. *Also called* shadow mask.

automatic degausser—An arrangement of degaussing coils mounted around a color picture tube. These coils are energized for only a short time when the receiver is turned on. They serve to demagnetize any parts of the picture tube that may have been affected by the earth's magnetic field.

background control—In color television, a potentiometer which is used as a means of controlling the dc level of a color signal at the input of a tricolor picture tube. The setting of this control determines the average (or background) illumination produced by the associated color phosphor. *Also called* drive control.

balanced modulator—*See* doubly balanced modulator.

bandpass-amplifier circuit—A stage designed to amplify uniformly signals of certain frequencies. In color television, the stage which amplifies the chrominance signal and its sidebands.

beam convergence—A condition produced in a three-gun color picture tube when the three electron beams meet or cross at a common point.

beam-positioning magnet—A magnet used with a tricolor picture tube to influence the direction of one of the electron beams so that it will have the proper spatial relationship with the other two beams.

black-and-white—*See* monochrome.

blue-lateral magnet—*See lateral-correction magnet.*

brightness—The attribute which makes an area appear to reflect or emit more or less light.

chroma—That quality which characterizes a color without reference to its brightness; that quality which embraces hue and saturation.

chromaticity—Quality, state, degree, measure of color.

chromaticity co-ordinate—The ratio of any one of the tristimulus values of a sample to the sum of the three tristimulus samples.

chromaticity diagram—A plane diagram formed when one of the three chromaticity co-ordinates is plotted against another. The most common chromaticity diagram is the CIE diagram plotted in rectangular coordinates. *See* Fig. 1-12.

chrominance cancellation—A cancellation of the brightness variations produced by the chrominance signal on the screen of a monochrome picture tube. Because of the persistence (or time-integration characteristic) of human vision, the brightness variations produced during the scanning of any one line are effectively cancelled by the brightness variations which occur during the scanning of the same line in the succeeding frame. In addition, a point which is made brighter during the scanning of a line in any field will lie directly above a point which is made darker during the scanning of the succeeding line in the same field. The brighter and darker points are not seen individually by the human eye because of space integration, and the illumination at one point on the screen is a value which is halfway between the brighter and darker values or the same as that which would be produced in the absence of the chrominance signal.

chrominance channel—In color television, a combination of circuits which have been designed to pass only those signals having to do with the reproduction of color.

chrominance signal—That portion of the composite color signal used to represent electrically the hues and saturation levels of the colors in a televised scene.

chrominance subcarrier—A signal having a specific frequency of 3.579545 megahertz used as carrier for the I and Q signals.

CIE—Abbreviation for "Commission Internationale de l'Eclairage." These are the initials of the official French name of the "International Commission on Illumination."

color-bar pattern—A test pattern consisting of a series of vertical bars, each having a specific hue, used to check the performance of a color television receiver.

color-burst signal—Approximately nine cycles of the chrominance subcarrier added to the rear porch of the horizontal blanking pedestal of the composite color signal and used in the color receiver as a phase reference for the 3.58-megahertz oscillator.

color-difference signal—A signal which results from the reduction of the amplitude of a color signal by an amount equal to the amplitude of the luminance signal. Color-difference signals are usually designated as $R - Y$, $B - Y$, and $G - Y$. In a sense, I and Q signals are also color-difference signals because they are formed when specific proportions of $R - Y$ and $B - Y$ color-difference signals have been combined.

color killer—A stage in a color receiver designed to prevent signals from passing through the chrominance channel during the reception of monochrome telecasts.

color-phase diagram—A vector diagram which denotes the phase difference between the color-burst signal and the chrominance signal for each of the three primary and three complementary colors. In many instances, this diagram is also used to designate vectorially the peak amplitude of the chrominance signal for each of these colors, and the polarities and peak amplitudes of the in-phase and quadrature portions required to form these chrominance signals.

color-picture signal—A signal which represents electrically the three color attributes (brightness, hue, and saturation) of a scene. A combination of the luminance and chrominance signals, excluding all blanking and synchronizing signals.

color purity—Freedom from mixture with white light or any colored light not used to produce the desired color.

color saturation—The degree to which white light is absent in a particular color. A fully saturated color is one which contains no white light. If 50 percent of the light intensity is due to the presence of white light, the color has a saturation of 50 percent.

color signal—In this book, a signal which varies in amplitude according to the intensity of the red, green, or blue light reflected by the televised scene.

complementary color—Every color has a complement. In the light additive process, the combination of any color with its complementary color will produce white.

composite color signal—The color-picture signal plus all blanking and synchronizing signals. The signal includes luminance and chrominance signals, vertical- and horizontal- sync pulses, vertical- and horizontal-blanking pulses, and the color-burst signal.

convergence—*See* dynamic convergence *and* static convergence.

convergence coil—One of the two coils associated with an electromagnet which is used to obtain dynamic beam convergence.

cross-hatch pattern—A series of horizontal and vertical lines produced on the screen of a television receiver by a pattern generator. This pattern can be used to check overall beam convergence and for making convergence adjustments.

cw reference signal—In color television, a sinusoidal signal used to control the conduction time of a synchronous demodulator. It must have the same frequency and phase as the original color subcarrier at the transmitter.

degaussing coil—A coil used to demagnetize any parts of the picture tube that may have been magnetized by the earth's magnetic field.

delay line—A specially constructed cable used in the luminance channel of a color receiver to provide a time delay to the luminance signal.

dichroic mirror—A special mirror through which all of the light frequencies pass except those of the color which the mirror is designed to reflect.

direct light—Light from a self-luminous object such as the sun or an incandescent lamp.

divided-carrier modulation—A process by which two signals can modulate two carriers having the same frequency, but having phase angles which differ by 90 degrees. The addition of the two modulator output signals will produce a resultant signal which has the same frequency as the carriers, but which will vary both in amplitude and phase in accordance with the variations in the amplitudes of the two modulating signals.

dominant wavelength—The single wavelength of light which predominates in any particular color.

dot pattern—Small dots of light produced on the phosphor screen of a picture tube by the signal from a dot, or pattern, generator. On the screen of a color receiver, there should be dot patterns of red, blue, and green light. If overall beam convergence has been obtained, the three dot patterns will combine to produce a white-dot pattern.

doubly balanced modulator—A modulator circuit in which two class-A amplifiers are supplied with modulating signals of equal amplitudes and opposite polarities and with carrier signals of equal amplitudes and opposite polarities. Carrier suppression takes place because a common plate circuit is used for the two amplifiers, and only the sidebands appear at the output.

drive control—*See* background control.

dynamic convergence—A term used to define the condition which exists when the three beams of a color picture tube converge at the aperture mask as they are deflected both vertically and horizontally. *Also see* vertical dynamic convergence *and* horizontal dynamic convergence.

electromagnetic convergence—The use of an electromagnetic field to obtain dynamic beam convergence.

electrostatic convergence principle—A principle used in early color receivers whereby beam convergence is obtained through the use of an electrostatic field.

equal-energy white—The light produced by a source which radiates equal energy at all visible wavelengths.

gray scale—White light of various intensities between black and white.

horizontal dynamic convergence—Convergence of the three electron beams at the aperature mask during the scanning of a horizontal line at the center of the picture tube.

horizontal parabola controls—*See* phase controls.

hue—The name of a color such as red, yellow, blue, or the like. Black, white, and values of gray are not considered as hues or as having hue.

illuminant C—The reference white of color television; light which most nearly matches average daylight.

indirect light—Light from an object which does not have self-luminous properties but which reflects light provided from another source.

in-phase portion (of the chrominance signal)—That portion of the chrominance signal having the same or opposite phase of the subcarrier modulated by the I signal. This portion of the chrominance signal may lead or lag the quadrature portion by 90 electrical degrees.

interleaving—Means placing between; as applied to the transmission of a composite color signal, the bands of energy of the chrominance signal are interleaved with or placed between the bands of energy of the luminance signal.

I signal—A signal which is formed by the combination of an $R - Y$ signal having a $+.74$ polarity together with a $B - Y$ signal having a $-.27$ polarity. One of the two signals used to modulate the chrominance subcarrier, the other being the Q signal.

keyed-rainbow generator—A color generator that produces a series of ten bars which vary in hue from yellow-orange on the left to green on the right of the screen of a color television receiver.

lateral-correction magnet—An auxiliary component of a three-gun picture tube which employs the magnetic convergence principle; it is used in conjunction with a set of pole pieces mounted on the blue gun as a means of positioning the blue beam in a horizontal direction so that proper convergence will be obtained.

luminance—The photometric quantity of light radiation.

luminance channel—In color television, the circuits which have been designed to pass only the luminance or monochrome signal.

luminance signal—A signal that varies in amplitude according to the luminance values of a televised scene. Part of the composite television signal.

magnetic-convergence principle—A principle whereby beam convergence is obtained through the use of magnetic fields.

matrix section—In color television, a combination of circuits to which signals having certain specifications are applied in order that signals having other specifications may be formed. The color-difference signals $R - Y$ and $B - Y$ are applied in a specific ratio to a matrix unit at the transmitter so that I and Q signals will be formed. The I, Q, and Y signals are applied in specific ratios to a matrix section in the color receiver so that the three color signals will be formed.

Maxwell triangle—A graph which defines the chromaticity values of a color in terms of three coordinates. *See* Fig. 1-11.

millimicron—A unit of length equal to one ten-millionth of a centimeter (10^{-7} cm).

monochrome—Having only one chromaticity, usually achromatic or black and white.

monochrome signal—Same as the luminance signal except that the term "monochrome" defines this signal as part of a transmission meant to be reproduced only in values of gray.

nonsaturated color—A color that is not pure; one that has been mixed with its complementary color or with white.

NTSC—Abbreviation for "National Television System Committee."

NTSC triangle—On a chromaticity diagram, a triangle which defines the gamut of color obtainable through the use of phosphors.

objective lens—In an optical or electronic system, the first lens through which the light rays pass.

phase controls—Controls used to adjust the phase of a voltage or current; specifically, in a color television receiver that employs the magnetic convergence principle, the three controls used to provide a means of varying the phases of the sinusoidal voltages which are applied at the horizontal scanning frequency to the coils of the magnetic convergence assembly. *Called* horizontal-parabola controls by some manufacturers.

phosphor dots—Minute particles of phosphor (phosphorus) on a viewing screen of a picture tube. In the tricolor picture tube, red, green, and blue phosphor dots are placed on the viewing screen in a pattern of dot triads—a phosphor dot of each color forms one-third of a triad.

phosphor stripes—The phosphors on the viewing screen of some in-line picture tubes are deposited in a stripe pattern. The green, red and blue phosphor stripes replace the phosphor dots used in conventional picture tubes.

polarity—The quality or condition inherent in a body which exhibits opposite or contrasted properties or powers in opposite or contrasted parts of directions; particular state (positive or negative).

pole pieces—Pieces of magnetic material positioned near the poles of a magnet to localize the field of magnetic force.

primary colors—A set of three colors from which all other colors may be regarded as derived; hence, any of a set of stimuli from which all colors may be produced by mixture. Each primary color must be different from the others, and a combination of two primaries must not be capable of producing a third. In color television, the three primary colors are red, blue, and green.

purity—*See* color purity.

purity magnet—A device consisting of two magnetic rings that is used in a color television receiver to produce a magnetic field which will alter the directions of the three electron beams so that each beam will strike only the correct color phosphor.

Q signal—A signal which is formed by the combination of $R-Y$ and $B-Y$ color-difference signals having positive polarities of .48 and .41, respectively. One of the two signals used to modulate the chrominance subcarrier, the other being the I signal.

quadrature—The state or condition of two related periodic functions or of two related points being separated by a quarter of a cycle or 90 electrical degrees.

quadrature amplifier—A stage used to supply two signals having the same frequency and having phase angles which differ by 90 electrical degrees.

quadrature portion (of the chrominance signal)—That portion of the chrominance signal having the same phase as, or the opposite phase of, the subcarrier modulated by the Q signal. This portion of the chrominance signal may lead or lag the in-phase portion by 90 electrical degrees.

reactance tube—A stage connected in parallel with the tank circuit of an oscillator so that the signal produced by the tube will either lead or lag the signal produced by the tank circuit. If the signal produced by the tube leads the signal produced by the tank circuit, the reactance tube is said to be inductive. If the signal produced by the tube lags the signal produced by the tank circuit, the reactance tube is said to be capacitive. The resultant oscillator signal will be a combination of the signal produced by the tube and the one produced by the tank circuit, and the phase (or frequency) of this resultant signal can be advanced or retarded when the control bias on the reactance tube is increased or decreased.

reference burst—*See* color-burst signal.

saturated color—A pure color; one that is not contaminated by white.

saturation—*See* color saturation.

secondary color—A color produced by combining any two primary colors in equal proportions. In the light additive process, the secondary colors are cyan, magenta, and yellow.

shadow mask—*See* aperture mask.

static convergence—Convergence of the three electron beams at the center of the aperture mask in a color picture tube. The term "static" applies to the theoretical paths which the beams would follow if no scanning forces were present.

subcarrier—A carrier conveyed by another carrier. A portion of the signal conveyed by a carrier which by itself performs the functions of a carrier.

synchronous demodulator—A demodulator which utilizes a reference signal having the same frequency as the carrier or subcarrier that is to be demodulated.

tilt controls—In a color television receiver which employes the magnetic convergence principle, the three controls used to tilt the vertical center rows in the three colored patterns produced on the viewing screen by the signal from a dot generator. Each control provides a means for adjustment of the amplitude of the sawtooth voltage which is applied at the vertical scanning frequency to one of the coils of an associated electromagnet, and the setting of each control determines the time during the vertical scanning period when the magnetic convergence force has the greatest effect on an associated electron beam.

triad (of dots)—A group of three different colored dots spaced so that a hypothetical line joining the dot centers would form an equilateral triangle.

tristimulus values—The amounts of each of the three primary colors that must be combined to establish a match with a given color sample.

tricolor camera—A television camera designed to separate the light reflected by a scene into three frequency groups which correspond to the light energies of the three primary colors. The camera transforms the intensity variations of each primary-color light into amplitude variations of an electrical signal.

tricolor picture tube—A picture tube which reproduces a scene in terms of the three primary colors.

ultor—An adjective used to identify the anode or element of a picture tube which is farthest removed from the cathode or to identify that anode to which the highest voltage is applied; also used to identify the voltage applied to this anode. The ultor anode is the second anode of the picture tube, and the ultor voltage is the voltage which is applied to the second anode.

vector—On a diagram, a line terminated in an arrowhead and used to designate the magnitude and direction of a force. In the field of electronics, vectors are often used to designate the amplitude and phase of a sinusoidal signal, or the amplitude and phase of a resultant signal formed by a combination of two or more sinusoidal signals. A complete discussion of vectors appears in Appendix C.

vertical-amplitude controls—A term used to define three of the controls in a color television receiver that employs the magnetic convergence principle. These controls provide a means by which the amplitude of the parabolic voltages applied at the vertical scanning frequency to the convergence coils may be adjusted.

vertical dynamic convergence — Convergence of the three electron beams at the aperture mask of a color picture tube during the scanning of each point along a vertical line at the center of the tube.

video-gain control—A control which provides a means for adjustment of the amplitude of a video signal. Two such controls are provided in the matrix section of some color television receivers so that the proper ratios between the amplitudes of the three color signals can be obtained. *Also called* drive control.

vsm alignment—An alignment technique, called video-sweep modulation, used to check the overall chroma frequency response of the video i-f section, video detector, and the chroma-bandpass amplifier.

white-dot pattern—*See* dot pattern.

Answers to Questions

Chapter 1

1. The eye will see the color yellow if the areas of red and green are small enough and if they are viewed from the proper distance.
2. The green region of the visible spectrum appears the brightest because the luminosity response of the average eye is peaked at the green region.
3. Desaturation is the opposite of saturation. A fully saturated color contains no white light; where, a desaturated color contains some amount of white light.
4. The colors on the line drawn from point E to any point on the spectrum locus are of the same hue but vary in degree of saturation. The closer the color is to point E, the less saturated it will be.
5. The three NTSC primaries are red, green, and blue. The three secondary colors are cyan, magenta, and yellow. Cyan is the combination of green and blue, magenta is the combination of red and blue, and yellow is the combination of red and green.

Chapter 2

1. A frequency, which is an odd multiple of one half the line frequency, was selected for the chrominance signal so that interleaving would take place.
2. The chrominance cancellation effect is a condition whereby the brightness variations produced on the screen of the picture tube by the chrominance signal apparently cancel each other.
3. The phase of the output sidebands will also change 180 degrees when the polarity of the modulating signal is reversed.
4. The resultant signal would be a sine wave having at any instant an amplitude equal to the algebraic sum of the values of the two output signals at that same instant, and the phase will be within the 90-degree separation of the two output signals and will favor the phase of the one having the greater amplitude.
5. The purpose of the color burst is to synchronize the frequency of the 3.58-MHz oscillator in the receiver with the frequency of the 3.58-MHz subcarrier used at the transmitter.

Chapter 3

1. When the scene is colorless or gray, the three outputs of the color camera will have equal amplitudes.
2. The signal for a fully saturated blue produces the darkest value of gray on the screen of a monochrome receiver.
3. The bandwidth is 1.5 MHz for the I signal, .5 for the Q signal, and 4.2 MHz for the Y signal.
4. For red, both I and Q are positive; for green, both are negative; and for blue, I is negative and Q is positive.
5. The color picture signal consists of luminance signal Y and the two chrominance signals (I and Q).

Chapter 4

1. Poor color or possibly complete lack of color on the screen.
2. The position of the sidebands is reversed from that occupied in the transmitted signal: upper sidebands are below the picture carrier; lower sidebands are above the picture carrier.
3. In a color receiver the sound-takeoff point must be ahead of the final i-f sound trap. In a monochrome receiver, the 4.5-MHz intercarrier sound is available in the detected video signal.

4. Signal level changes at the video detector will produce color changes in the final picture, and the result will be incorrect colors.

Chapter 5

1. The delay line provides a time delay for the luminance signal so that it will arrive at the picture tube at the same time as the chrominance signal.
2. The agc tends to maintain a constant signal level at the detector. In a color receiver, a change in signal level can produce a color change on the screen. On a black-and-white receiver this level change is noticed only as a difference in gray level.
3. No, differences are not necessary. The same basic circuit may be used in both color and monochrome receivers.
4. If the ultor voltage were allowed to change, the amount of deflection at the edges of the screen would also change. This would cause the picture size and convergence to change as the brightness varies.
5. A thermistor is connected in series with the secondary winding of the power transformer and the rectifier circuit. The degaussing coil, in series with a voltage-dependent resistor, is connected across the thermistor. When the set is first turned on, the thermistor is cold and most of the power-supply current flows through the degaussing coil and the voltage-dependent resistor. As the thermistor warms due to the small current through it, its resistance decreases and passes an increasing portion of the power-supply current. As less current flows through the voltage-dependent resistor, its resistance increases and restricts the current through the degaussing coil to a negligible amount.

Chapter 6

1. To separate the chrominance signal from the composite color signal.
2. A horizontal keying pulse turns the chroma-bandpass amplifier off during horizontal-retrace time.
3. Both signals are 3.58 MHz. They differ only in phase relationship.
4. The color killer turns the chroma-bandpass amplifier off during a monochrome telecast.
5. The color burst is used as a reference signal to control the frequency and phase of the cw reference oscillator.

6. The "hue" or "tint" control changes the phase of the cw signal and this affects the hue of the colors which appear in the reproduced scene.

Chapter 7

1. Both amplitude and phase modulation are present in the chrominance signal.
2. A synchronous demodulator is used to sample portions of the chrominance signal.
3. The two output signals represent the original I and Q signals developed at the transmitter.
4. The 3.58-MHz component is filtered out at the plates or outputs of the demodulators.
5. The polarity of the signals at the demodulator outputs will be reversed.
6. The phase relationship between these signals is:
   (a) 90 degrees.
   (b) 90 degrees.
   (c) 90 degrees.

Chapter 8

1. Zero.
2. $R - Y$, $B - Y$ and $G - Y$.
3. The polarity of the luminance signal applied to the cathodes of the color picture tube must be such that the sync pulses are the most positive part of the signal.
4. To restore the dc reference level to the color signals.
5. Yes.
6. No.

Chapter 9

1. The screen is composed of green, red, and blue phosphor dots arranged in triads or phosphor stripes.
2. The blue gun will be positioned above the red and green guns.
3. (c) Convergence.
4. False. Each beam-positioning magnet affects only one beam.
5. Static and dynamic.
6. To reduce the effect of stray magnetic fields on the three electron beams.
7. Adjustment of the purity magnet.
8. A blue-lateral correction magnet.
9. Dynamic convergence.
10. (a) Sawtooth.
11. The convergence signals are obtained from the horizontal and vertical sweep circuits.

Chapter 10

1. Three beam-positioning magnets and one blue-lateral correction magnet.

2. Demagnetizing the color picture tube.
3. To correct for the differences in the efficiencies of the three phosphors used, and also to reproduce a proper gray scale.
4. A low-power microscope can be used to observe the beam landing on individual phosphor dots.
5. Sixteen controls.
6. (a) Four controls.
   (b) Six controls.
   (c) Six controls.

Chapter 11

1. Check the reception of a monochrome signal.
2. In the color sections.
3. Diagonal stripes of variegated colors will appear on the screen.
4. The trouble will be in the antenna, lead-in, or improper antenna orientation.
5. Colors will be desaturated.
6. The oscillator phase is incorrect.
7. Yes. The oscillator control or the phase-detector tubes.
8. The picture tube should be demagnetized, or degaussed.
9. Diagonal color stripes will be stabilized on the screen.
10. No. The i-f bandpass requirements of a color television receiver are considerably wider than those of a monochrome receiver.

Index